Suomen Muinaismuistoyhdistys ry – Finska Fornminnesföreningen rf
The Finnish Antiquarian Society

Materiality and Objects:

Multi-disciplinary Approaches to Archaeological Material and Contexts

Proceedings of the Finnish-Russian Archaeological Symposium

Hämeenlinna, 10-12 October, 2019

Editors:
Petri Halinen & Jouni Taivainen

ISKOS 24

Helsinki 2021

Editors, volume 24: Petri Halinen & Jouni Taivainen

English language revision: Grano Oy Translation Services
and Lisa Muszynski (Mika Lavento's article)

Cover design: Mikael E. T. Manninen

Cover photo: Museovirasto/Jouni Marjamäki

Layout: Mikael E. T. Manninen

Museovirasto
Museiverket
Finnish Heritage Agency

VERTAISARVIOITU
KOLLEGIALT GRANSKAD
PEER-REVIEWED
www.tsv.fi/tunnus

Printed in Finland by Grano Oy, Helsinki 2021.

ISBN 978-952-6655-28-4 (print)
ISBN 978-952-6655-29-1 (pdf)
ISSN 0355-3108

Contents

Preface

Some years ago, I was contacted by the Cultural Affairs Officer of the Consulate General of Finland in St. Petersburg. I still clearly remember where I was when I received the call; visiting the pioneer museum of Miehikkälä and negotiating its transfer to Hämeenlinna. Indeed, the purpose of that call I received was to bring Finnish and Russian archaeologists back together to reflect on common issues from a scientific perspective.

Professor Kirpichnikov, Dr. Nosov and Dr. Saksa, all of whom are known in Finland, are the founding fathers of the concept of a modern Finnish-Russian archaeological symposium. They all played a key role in the initial stages of the history of the symposium back in the 1970s. In the same decade, the Finns involved in this matter were Professor of Archaeology Carl-Fredrik Meinander from the University of Helsinki, Head of Department Torsten Edgren from the National Board of Antiquities, and Professor of Archeology Unto Salo from the University of Turku. As some of us remember, the first Finnish-Russian archaeological symposium was held in 1976.

Both of the topics discussed that day in Miehikkälä have been realised. The museum has been moved to Hämeenlinna, where it is now situated near Häme Castle. But for us, the most important thing is that the Finnish-Russian archaeological symposium has been reorganized, and in the beginning of October 2019 we had the possibility of opening the third modern Finnish-Russian archaeological symposium in Hämeenlinna. We met in Helsinki for the first time in 2014, then again in Novgorod, and now in Hämeenlinna.

Last time we shared these thoughts in Russia in Novgorod two years ago, in 2017. As a result of a relatively short preparation period we obtained Hämeenlinna symposium, which now returns to its roots, but I hope this will create entirely new scientific cooperation between Russia and Finland. I am glad that all Finnish universities involved in archaeology research and teaching are among the participants, and that there is also a large number of Russian archaeologists attending this event.

When it comes to Finnish archaeologists, a complete generational change has occurred in the intervening decades. Now there are only a few people left who even remember the early stages in some way. Most of us were young and full of hope for the future in the late 1970s. Perhaps we remember the early stages of this cooperation. While some of us have remained close to science all these years, I instead lost sight of it and became increasingly involved in administrative matters. The young generation is now in a key position to create this archaeological cooperation in spite of the difficult global political and economic situation.

In this context, also, I would like to thank Professor Vladimir Lapshin (chair), Evgeniia Tkach and Dr. Alexander Saksa on the initiative of the symposium, as well as Dr. Petri Halinen, Dr. Päivi Maaranen and Mervi Suhonen for arranging the symposium. I would also like to thank the Finnish Antiquarian Society for the cooperation with the organisation and the City of Hämeenlinna for the warm hosting of the symposium.

Generations have passed and will continue to pass also in the Finnish-Russian co-operation. The activities in the coming years will be observed by several retired persons who remain active, such as Dr. Torsten Edgren, Dr. Pirjo Uino and Dr. Paula Purhonen, to name but a few. The Hämeenlinna symposium was the last time I hosted this kind of co-operation. In the beginning of 2020 I retired and have become a researcher again. For me personally the reinitiating of the cooperation was important. That is why I will continue to follow what you do. Generational change is a good thing, after all. It is a necessity in this life cycle. The entry of young researchers to this joint action is an indication of the growing interest in the prehistoric cultures of the Finnish-Russian region.

To our grief, the 'grand old men' of the Finnish-Russian archaeology symposia have passed away. Professor Evgeniy N. Nosov (1949–2019) passed away just before the Hämeenlinna symposium. Shortly after our symposium in October 2019, another Russian 'grand old man' passed away. Professor Anatoliy N. Kirpichnikov (1929–2020) died in the beginning of 2020. Both of them had been involved in the entire period of cooperation. Professor Unto Salo, who was involved in the cooperation in the 1970s and 1980s, died in autumn 2020. We warmly remember them all for valuably maintaining and promoting the cooperation.

Through all these years, one visible result of the Finnish-Russian meetings has been the publication of articles. It has been, and hopefully will continue to be, a tangible imprint regarding the cooperation, which can still be evaluated after all these decades. I want to thank the Chief Intendant, Dr. Petri Halinen, for editing the publication, and for giving me the honour of writing 'as my last job in office' this greeting. I also want to thank Intendant Jouni Taivainen for helping to edit this publication, and several archaeologists in Finland and abroad for participating in the peer review process. I am grateful to all Russian and Finnish researchers who have submitted their presentations for publication for the sake of everyone who is interested in Finnish and Russian archaeological research.

Juhani Kostet
Docent, Director general emeritus

For the Reader

The 16th Finnish-Russian archaeological symposium was held in Hämeenlinna from 9 to 10 October 2019, with the excursion day following the symposium on 11 October. The theme of the symposium was *Materiality and Objects: Multi-disciplinary Approaches to Archaeological Material and Contexts*. The theme was approached through the subjects of multidisciplinarity and interdisciplinarity. The presentations dealt with topics ranging from the Stone Age to the Middle Ages. A total of 16 presentations and two posters were held during the two days of the symposium.

Häme Castle is a medieval castle with a seminar room for conferences and symposia similar to this one. Sixty-one people participated in the symposium and the excursion, which was directed to Häme and Uusimaa. Guided visits to the medieval Holy Cross Church in Hattula, the ancient hillfort of Rapola in Valkeakoski, the Iron Age cemetery in Janakkala and the Mesolithic site of Brunaberget in Vantaa were included in the excursion.

Contributions to the symposium were selected in Finland through an open call for papers. In Russia, the Russian members of the organising committee selected or invited speakers within the Institute for the History of Material Culture, and only a few Russian participants outside this institution joined the symposium.

Iskos, the series of publications of the Finnish Antiquarian society, was selected as the publishing forum. Several previously published Finnish-Russian archaeological symposia have also been published in Iskos. The present volume includes 10 papers, which had been held as presentations or posters. All the articles have been peer-reviewed by two anonymous referees.

The articles have been arranged in chronological order, from the Stone Age to the historical period. This was done for practical reasons.

The authors were responsible for translating their articles into English. The language check was done mainly by Grano Ltd. Mika Lavento's language check was done by Lisa Muszynski.

The editors wish to thank Mr Mikael E. T. Manninen for designing the cover and layout of the volume. We want to also thank the 21 anonymous reviewers who gave many good suggestions for improvement regarding the texts. We thank the Finnish Antiquarian Society for the financing of the volume's layout and printing it. The Finnish Heritage Agency, Department of Cultural Environment Services, enabled the editing and language check for the book.

In Helsinki, 17 December 2021
Petri Halinen & Jouni Taivainen

Periodization and Chronology of the Eneolithic in the Upper and Middle Kama Regions

Aleksandr A. Vybornov, Eugeniya L. Lychagina, Marianna A. Kulkova & Andrey F. Melnichuk

Aleksandr A. Vybornov, Samara State University of Social Sciences and Education, M. Gorykogo st., 65/67, Samara, 443099, Russia, email: vibornov_kin@mail.ru
Eugeniya L. Lychagina, Perm State University, Bukireva Str., 15, Perm, 614990; Perm State Humanitarian Pedagogical University, Sibirskaya street, 24, Perm, 614990 Russia, email: LychaginaE@mail.ru
Marianna A. Kulkova, Herzen State Pedagogical University, 191186 St.Petersburg, nab. Moyki,48/12, Russia, email: kulkova@mail.ru
Andrey F. Melnichuk, Perm State University, Bukireva Str., 15, Perm, 614990, Russia

Abstract

The territory of the Upper and Middle Kama regions is important for study on account of the Turbino sites that have been open there. Artifacts of this type have also been found in Finland. Subsequently, there have been several topics of discussion. Some archaeologists considered the Garino cultural complexes to be from an earlier period, while others supposed that they had appeared simultaneously with the Bor cultural type. The single radiocarbon dates were only obtained for the Garino culture. Therefore, the periodization and chronology of the Eneolithic in the Kama region have not been developed. In recent years, new archaeological sites have been excavated and a series of radiocarbon dates have been obtained for both types of cultures. This allowed us to work out the periodization and chronology of the Eneolithic of the Upper and Middle Kama regions in more detail. The chronological framework of the Garino culture was determined from 4500 BP to 3500 BP (ca. 3500-1600 calBC). The sites of the Bor cultural type were occupied from 4200 BP to 3900 BP (ca. 3000-2200 calBC). Thus, the chronological priority of the Garino culture is established. According to the typological and chronological framework, two stages in the development of the Garino culture were elaborated. The coexistence of the Garino and Bor complexes had a place at the end of the early stage of the Garino culture.

1 Introduction

The study of the Eneolithic sites in the Upper and Middle Kama regions began in the 1930s (Shmidt 1935; Prokoshev 1935). As a result of large-scale excavations in the 1940s and 1950s, O.N. Bader and his students created a considerable source base for this period. On this basis, the researcher identified the Turbino culture. He combined the settlements of the Garino and Bor type with the Turbino burial ground into this culture (Bader 1963; 1964). O.N. Bader attributed all materials to the Bronze Age back to the 2nd millennium BC (Bader 1961a; 1961b). The researcher identified the early (Garino) and late (Bor) stages as part of the Turbino culture. The chronological framework of the first stage was determined as being between the 20th century BC and the 14th century BC, with the second being in the 14th-12th centuries BC (Bader 1963: 25-30).

In the 1970s and 1980s, as a result of studying the chemical composition of the

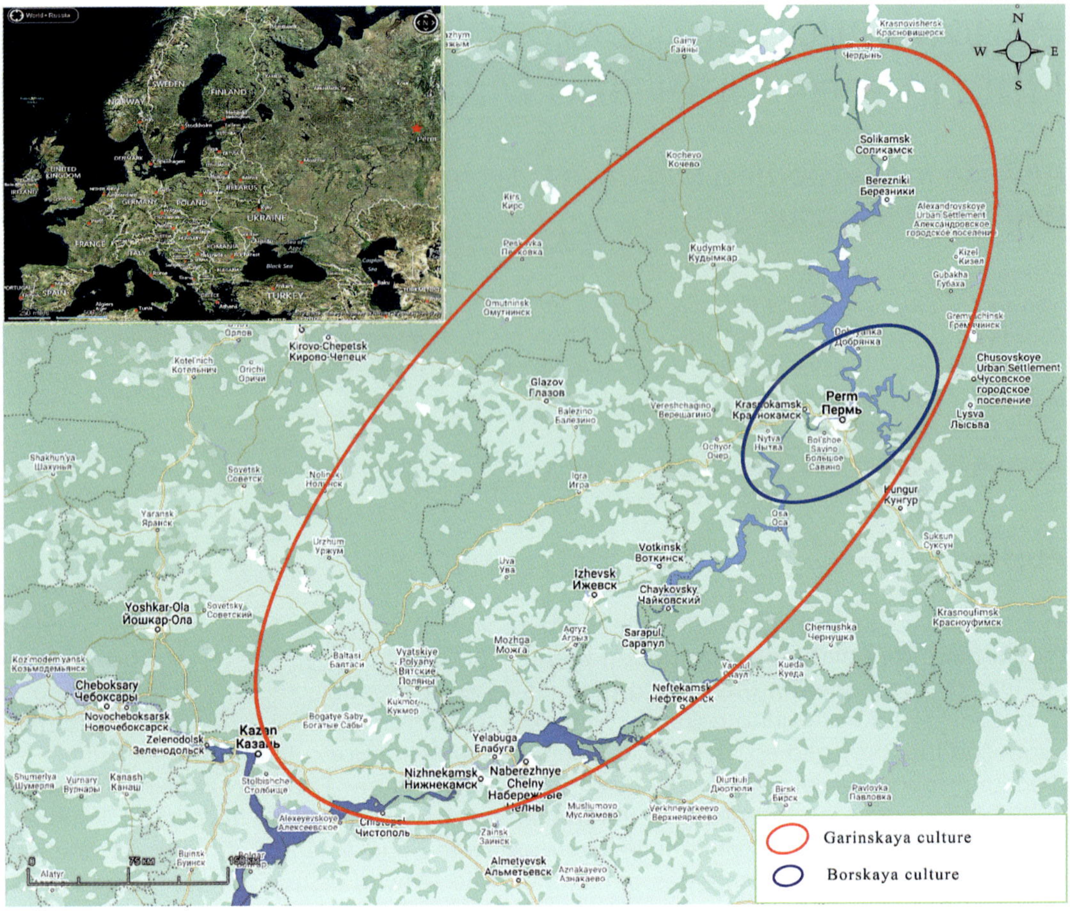

metal artifacts from the Turbino burial ground and settlements, some archaeologists made conclusions about their significant differences. In order to make a distinction between the materials of burial grounds of the Turbino type and settlement complexes, the term "Garino-Bor culture" was proposed for the latter (Chernykh 1970). As a consequence, the complexes of the Turbino burial ground were attributed to the Bronze Age, and the materials of the Garino and Bor settlements to the Eneolithic (Chernykh & Kuzminykh 1989). The authors of the article agree with the researchers' assertion that the Eneolithic is characterized by tools made from copper, whereas the Bronze Age began with the appearance of copper alloys (Ryndina & Degtyareva 2002). Further research allowed scientists to substantiate the independent status of the Garino culture and distinguish the Bor culture as a special type (Melnichuk 1990; Nagovitsyn 1990). However, there remained questions about the genesis and periodization of these cultures, as well as the chronology. This was due to the lack of stratigraphic data and a representative series of radiocarbon dates. This paper aims to advance the resolution of these contentious questions.

2 Garino culture

The territory of the Garino culture's distribution covers the entire Kama River basin up to the River Ik in the south. The greatest concentration of sites is in the estuary of the Chusovaya River near the cities of Perm, Osa, Okhansk, Tchaykovsky and Sarapul. More than 20 out of 100 reference sites were investigated, located on pine-forest terraces over a floodplain with a height of 4-10 m. Temporary seasonal sites (Bor II, Borovoe ozero IV, etc.) were studied. At the same time, permanent settlements with a large number of dwellings were investigated: Krasnoe Plotbishche (30), Bor I (24), Starushka I (12), Nepryakha VII (27), Borovoe ozero II (12), etc. The number of house pits is smaller in the sites of the Northern Kama region (Chashkinskoe ozero II, Vasyukovo II). A characteristic feature of early Garino construction is square pit-houses connected by passages and dug into the ground to a depth of 1 m. Their surface area varies from 40 to 100 m² (6 x 7 m; 8 x 8 m, etc.). Rectangular single dwellings were more typical on the Garino sites of the late stage.

Garino ceramics are characterized by porosity. O.N. Bader suggested that this is the result of the presence of pine bark admixture in the clay, burnt away during firing (Bader 1961a: 28). In the south of the Kama region (Simonikha II, Krasnoe Plotbishche), Garino culture dwellings were found in loamy soils, and the ceramics found have an admixture of the crushed shell. Apparently, it was dissolved in sandy soils (Denisov & Melnichuk 2014: 50). This is confirmed by the recent analysis of the molding mass of the Garino culture's ceramics (Batueva & Lychagina 2018).

The Garino vessels have a cylindrical neck with a restricted or straight mouth and a rounded bottom. In addition to the rounded and straight rims, Г- and T-shaped rims are present. These were decorated with imprints of a toothed stamp, and less often with short oval impressions. Sometimes the pattern is located on the inside of the rim. A characteristic feature is the treatment of the inner surface of the vessels with a toothed instrument, which resulted in the formation of scratches (Fig. 1).

The ornament covers almost the whole surface of the vessel walls in horizontal and vertical zones. The vast majority of the pottery was decorated with a comb tool, which can be divided into prints of a long stamp and prints of a short oval. The second variety is more characteristic of the southern Garino sites of the Middle Kama region. Ornamental compositions are represented by belts of oblique or horizontal rows of imprints of a long or short stamp, a horizontal or vertical zigzag pattern, and oblique lattice. There is a motif in the form of a "stepping comb" (Fig. 2). Interestingly, "stepping" prints are more common on the pottery of the Upper Kama region than on the sites of the Middle Kama region. Geometric compositions in the form of rhombuses and triangles are present. These motifs resemble patterns of the Ayatskaya type of Trans-Ural origin. Similar ceramics were found in the dwellings of the Bor I settlement and others.

Flakes and spalls served as blanks for making tools. Blades are practically absent. Flint slabs were used for toolmaking. Among the tools are end scrapers and round scrapers, sometimes with a cut blade. The knives are leaf-shaped, rectangular, or curved, and sometimes have a "button-like" shape on the handle. Borers are less common. The arrowhead forms include leaf-shaped, willow-leaf-shaped and triangular ones with truncated or notched bases, including also pentagonal specimens (Fig. 3). Woodworking tools are present in the form of polished axes, adzes, and chisels. There are also grooved hammer stones. Stone sinkers for fishing nets confirm the presence of net fishing. Adornments are represented by oval or round slate pendants, and animals are represented by flint figurines.

Traces of metallurgy were found in many sites (Bor I, Boytsovo VI, Vystelishna, Starushka). A structure used for copper smelting was discovered in the Krasnoe Plotbishche settlement, some fragments of clay casting moulds were found in the settlement of Zaosinovo I, and more than 600 pieces of copper ore were found in a dwelling at the Sauz I settlement (Vybornov et al. 1984). In addition, various copper products were discovered: awls, leaf-shaped knives, and spiral

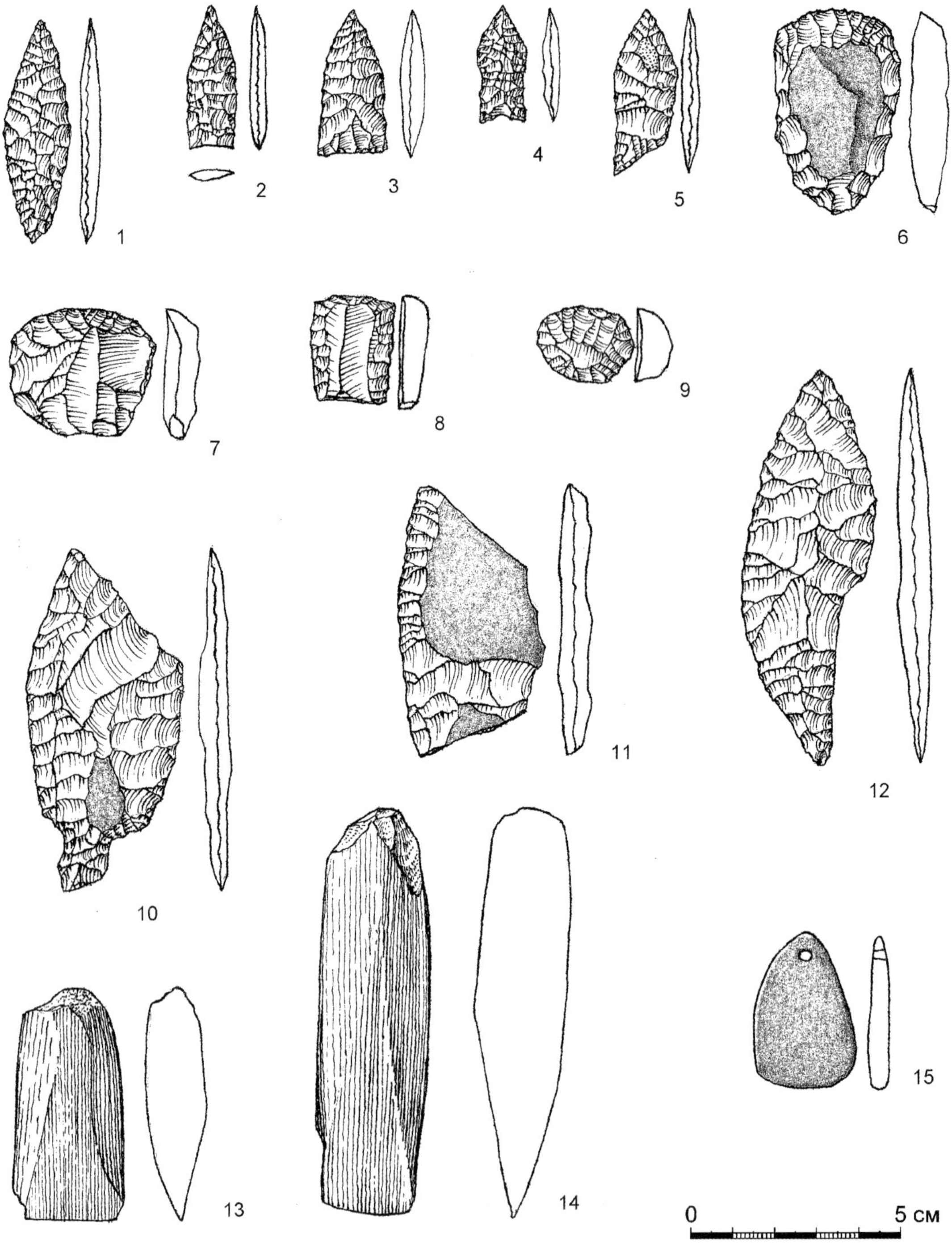

Figure 3. Stone tools of the Garino culture.

Three elongated rectangular structures were studied at the Bor V settlement, three at Bor III and one at Malo Borovoe ozero. The largest of them (36 x 6 m) was excavated at the Borovoe ozero VI settlement.

Borino ceramics are represented by vessels with open pot or bowl shapes, straight walls and a rounded bottom. They are porous because of the dissolution of shell inclusions added to the clay paste. The inner side of the vessels is scraped with a toothed instrument. The rims are slightly bent outward and slanting in cross-section. Ornamentation was sometimes applied to the top and the inside of the rim (Fig. 4), though it does not cover the entire surface of the vessel. The decorations were applied with a comb tool (80-90%). Unlike with the Garino type, impressions of short and medium stamps prevail. Another 10-15% of the vessels are decorated with pit-like impressions and about 10% with a smooth stamp. The ornamental compositions of the Bor vessels are simple: horizontal and oblique belts, and vertical and horizontal zigzags (Fig. 5). A characteristic feature of Bor ceramics is the lack of a "stepping comb" motif.

The stone industry of the Bor complexes is characterized by blade technology. In contrast to the stone industry of the Garino culture, flint slabs were hardly used as raw material. The percentage of tools with long and wide blades is quite high. Knives with straight edges have been made with large blades with an edge retouch on the long sides. End scrapers on massive spalls have a steep retouch on the longitudinal sides. The arrowheads are leaf-shaped, lanceolate, almond-shaped, and tanged, and they come with a truncated base. Axes, adzes, chisels, sinkers, and hammer stones are present, and the adornments include slate pendants (Fig. 6).

Traces of metalworking were only found in dwelling 2 of the Bor V settlement, where an awl and plate, two pouring funnels and three drops were found. A spectral analysis performed by E.N. Chernykh showed that, in terms of metal composition, these products are similar to the copper objects from the late Garino settlements of Basenkiy Borok and Vystelishna.

The genesis of the Bor culture is an issue of debate. According to O.N. Bader, the formation of the Bor culture was influenced considerably by elements of the Novoilinsk culture that appeared in the Chusovskoye Kama region in the late Garino stage (Bader 1961a: 184). Other researchers emphasise the local originality of the Bor culture in the Chusovskoye Kama region. There is also a school of thought that states that the Bor culture was formed based on the Novoilinsk culture instead of being based on Garino culture sites (Melnichuk 1990; Nagovitsyn 1990).

4 Periodization and chronology of the Garino and Bor complexes

While working on questions of periodization and chronology of the Eneolithic materials, O.N. Bader relied on archaeological data. Based on them, he believed that the sites of the Garino culture were from an early stage. On their basis, later Bor complexes were formed under the influence of the Novoilinskaya culture. The researcher noted the Neolithic appearance of materials of the Novoilinskaya type, but the discovery of copper objects there made him favour the main hypothesis. In the absence of radiocarbon dates, the chronological framework of these cultures was determined rather approximately.

In the 1990s, researchers referred again to these issues. They focused on the fact that features of the Kama Neolithic culture are traced in the Bor materials. These features include pot or bowl-shaped vessels, decoration patterns made with large oval shallow pits, a small percentage of arrowheads, and the shape and size of dwellings. Thus, many of the material culture components of the Bor cultural type of sites seemed to the researchers to be more archaic than the Garino ones. As for chronology, the Garino culture was dated to between the second half of the 3rd and the first half of the 2nd millennium BC

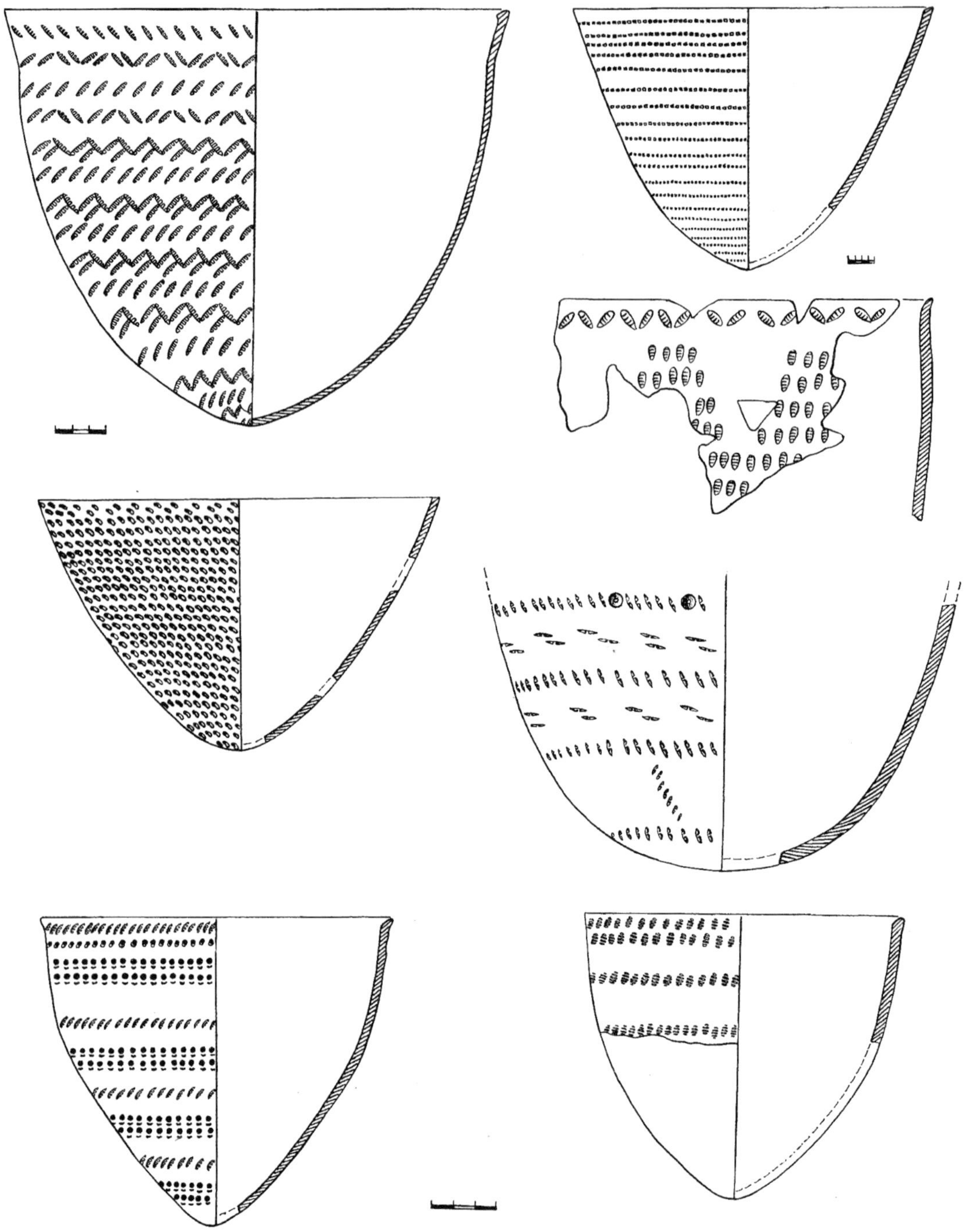

Figure 4. Ceramics of the Bor cultural type.

Figure 5. Ceramics of the Bor cultural type.

in the preliminary plan. A radiocarbon date was used to date the lower boundary, and the rest of the interval was dated by analogy. Researchers believed that the sites of the Bor culture existed for a shorter period and correlated with the early Garino ones (Nagovitsyn 1990; Melnichuk 1990).

Since 2007, the research team has carried out targeted work to create a base of radiocarbon dates of the Eneolithic in the Kama region (Melnichuk 2009; 2013; Lychagina & Vybornov 2009; Lychagina 2011; Vybornov et al. 2019). This made it possible to establish a quite reliable chronological framework of the Novoilinsk culture: from the last quarter of the 4th to the middle of the 3rd millennium BC. A comprehensive analysis of all materials of the Novoilinsk type did not confirm its belonging to metallurgical cultures (Denisov & Melnichuk 2014; Vybornov et al. 2019).

The analyses for the Garino and Bor settlements were obtained from charcoal and organics in the pottery matrix. The radiocarbon dates on charcoal were obtained for sites with a single cultural layer for this region. This is well correlated to archaeological periodization. Most of the dates on the organics of ceramics are reliable, and this was supported by other investigations on Neolithic and Eneolithic sites. For example, the dates on the organics of ceramics from the Baybek site in the Northern Cis-Caspian region are 6920±120 BP (SPb-1053) and 6925±120 BP (SPb-1716), coinciding with the dates on charcoal (6986±44 BP (Ua-50260) and 6948±120 BP (SPb-1713)). The date on the organics of ceramics from the Oroshaemoe site in the Lower Volga basin (5890±120 BP (SPb-1729)) is the same as the date on animal bones (5806±26 BP (UGAMS-23059)). Then

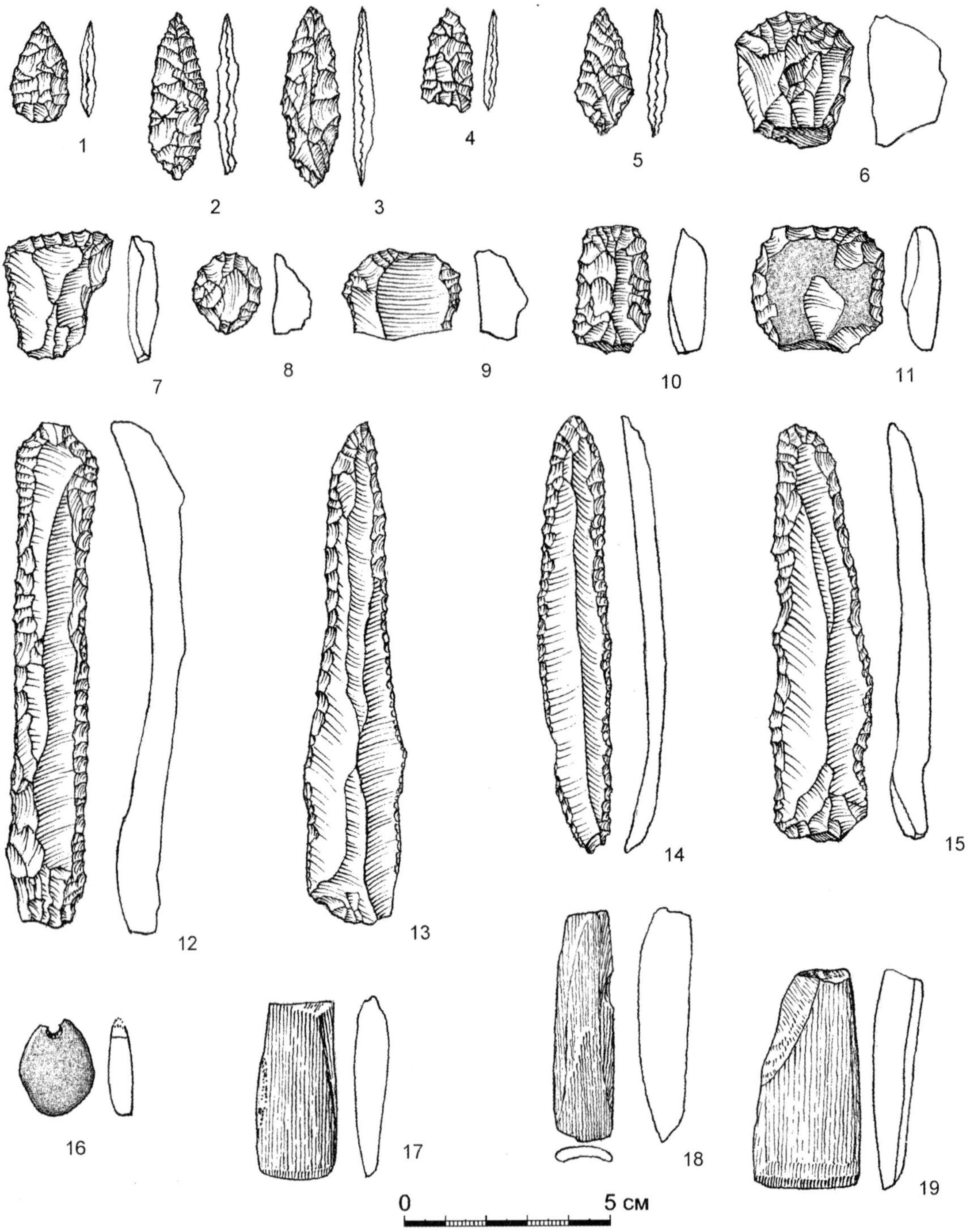

Figure 6. Stone tools of the Bor cultural type.

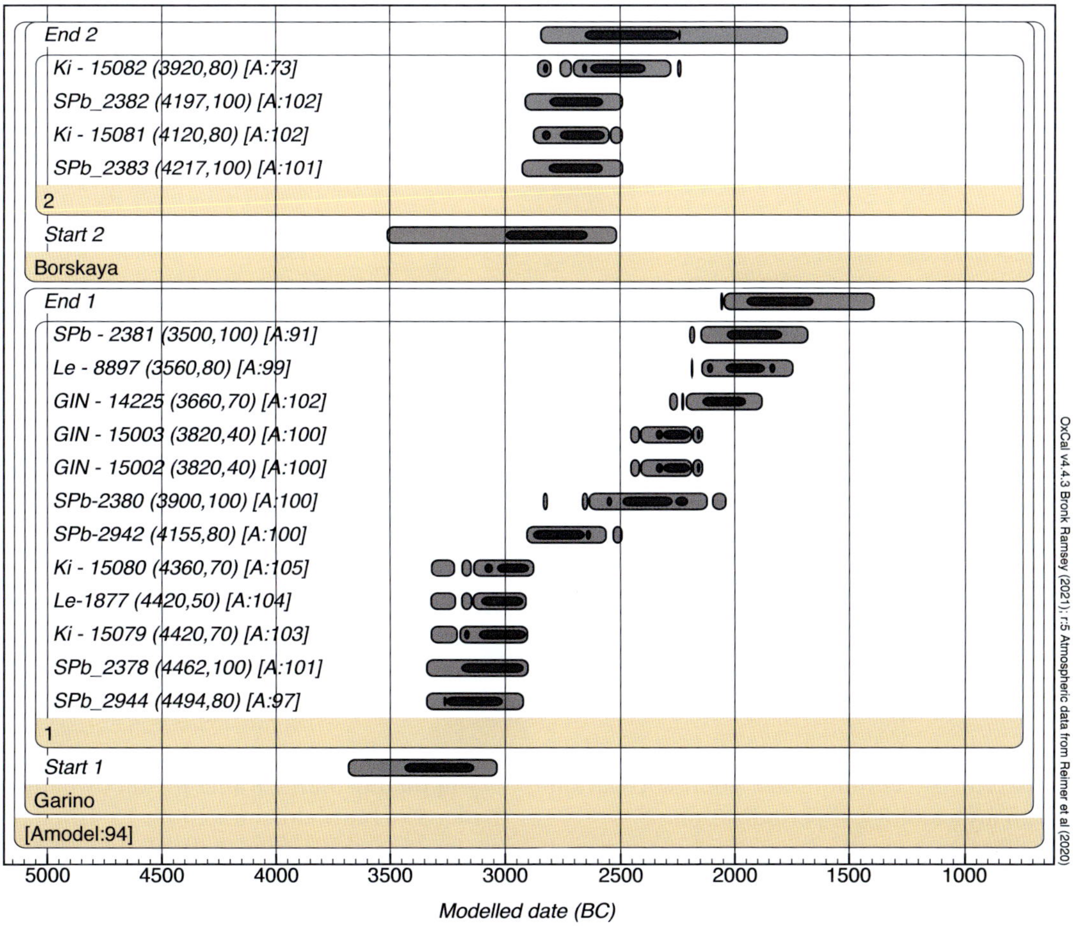

there are the dates for the Kalmikovka 1 site located in the forest steppe zone of the Volga basin. The date on the organics of ceramics is 5950±120 BP (SPb-1759). This is the same as the date on the animal bones: 5989±70 BP (SPb-1876). The date on Neolithic ceramics (7010±80 BP (Ki-15915)) from the Koksharovskiy Kholm site in the forest zone of the Ural region coincides with the date on charcoal (7050±180 BP (Le-7883)). The same situation exists on the Zamostye II site in the Upper Volga region. The dates on the ceramics of the Early Neolithic are 6830±80 BP (Ki-15533) and 6680±80 BP (Ki-15434). The dates on the charred crusts are 6834±63 BP (Ua-48463) and 6650±45 BP (Ua-37101). The Khvalynsk culture on the Kara-Huduk site located in the Northern Cis-Caspian region has two dates on the organics from ceramics: 5950±80 BP (Ki-14912) and 5820±80 BP (Ki-14911), while the dates on the animal bones are 5900±100 BP (SPb-2365) and 5854±60 BP (SPb-2338). For the Kama Neolithic culture of the Upper Kama region the dates on ceramics and charcoal are 5880±100 BP (SPb-897) and 5850±70 BP (GIN-15447), respectively. Specialists note that dating the ceramics with shell inclusions could yield the wrong date in some cases (Meadows, 2020). This type of ceramic composition gives a reservoir effect in dating. The Garino and Bor ceramics were made from clay paste without shell inclusions. The radiocarbon dates on organics of

Nº	Site	Index	Age (BP)	Age, calBC (2 s)	Material
			GARINO CULTURE		
1	Nepryakha IV	Le - 1877	4420±50	3350-2900	Charcoal
2	Bor I	SPb-2378	4462±100	3490 - 2901	Pottery carbon
3	Lake Borovoe II	Ki - 15079	4420±70	3340-2900	Pottery carbon
4	Lake Borovoe III	Ki - 15080	4360±70	3340-2870	Pottery carbon
5	Bor I	SPb-2380	3900±100	2637-2121	Pottery carbon
6	Lake Chashkinskoye IX	SPb-2944	4494±80	3484-2925	Pottery carbon
7	Lake Chashkinskoye IIIa	SPb-2942	4155±80	2904-2496	Pottery carbon
8	Lake Chashkinskoye IIIa	GIN - 15002	3820 ±40	2370-2190	Charcoal
9	Lake Chashkinskoye IIIa	GIN - 15003	3820 ±40	2370-2190	Charcoal
10	Novoil'inskoe III	GIN - 14225	3660±70	2300-1750	Charcoal
11	Novoil'inskoe III	Le - 8897	3560±80	2140-1690	Charcoal
12	Krasnoe plotbicshe	SPb-2381	3500±100	2061-1608	Pottery carbon
			BOR CULTURE		
13	Lake Borovoe VI	SPb-2383	4217±100	3036-2562	Pottery carbon
14	Lake Borovoe IV	Ki - 15081	4120±80	2890-2480	Pottery carbon
15	Bor V	SPb-2382	4197±100	3022-2550	Pottery carbon
16	Bor III	Ki - 15082	3920±80	2620-2190	Pottery carbon

Table 1. Radiocarbon dates of the Garino and Bor cultures.

this type of ceramics have good correlation with archaeological periodization.

To date, four ^{14}C dates have been received for the Bor sites (Table 1). Three of them have a rather close time interval, which indicates the validity of the obtained data. Thus, new data suggest the existence of the Bor culture in the Chusovskoye Kama region between the last quarter of the 3rd and the beginning of the 2nd millennium BC. The fact that the typologically earliest complex of the Borovoe ozero VI settlement received the earliest date also argues in favour of the validity of the dates. Judging by the dates, the upper boundary of the Novoilinskaya culture and the lower boundary of the Bor culture have a gap of about 400 years. This casts doubt on the chronological sequence of these cultures. To specify the periodization of the Bor sites, an additional comparative analysis of the materials of the earliest (Borovoe ozero VI) and latest (Bor III) sites is required.

As for the chronological frameworks of the Garino culture, its definition was based on the presence of Ayatskaya type of pottery in the Eneolithic Kama region dwellings (Bader 1961a; Melnichuk 2009: 15-16), dating back to the second half of the 3rd millennium BC. (Chairkina 2005: 289). It is permissible to date the beginning of the late Garino period based on the copper lunulas from the sites of Starushka, Vystelishna and Ust-Pal, an analogy of which is presented in the West Siberian dated complexes of the end of the 3rd millennium BC (Koksharov 2012: 32-34).

For the Garino culture, 12 radiocarbon dates from eight sites were obtained (Table 1). They fit into the interval from the middle of the 3rd to the middle of the 2nd millennium BC. Most dates of the first half of the 2nd millennium BC were obtained on charcoal (Table 1, 8-11), while most of the dates of the second half of the 3rd millennium BC were obtained on organics in the ceramic matrix (Table 1, 2-7). There is no inclusion of shells in the clay pastes of the fragments used for the analysis (the shells were probably dissolved due to being in sandy soils). It is noteworthy that the two dates on charcoal from the Chashkinskoe ozero IIIa site are 300 years younger than the dates for organics from ceramics (Table 1, 7-9). If that is not taken into account, then in the series of dates for the Garino culture two groups are formed: 4460-4360 BP (3500-2870 calBC) and 3900-3500 BP (2637-1600 calBC). It is interesting to note that the dates for the Borino type fit into the interval of 4200-3900 BP (3022-2200 calBC). At the same time, the earliest dates on organic matter in ceramic matrices coincide with the dates on charcoal. In addition, the earliest date corresponds to the earliest typological complex of the Bor I site, which confirms its validity. A similar situation exists with the latest date. It fits well in a series of data on charcoal. The earliest dates match with the latest data for the Novoilinsk culture. The chronological coexistence of the Garino and Bor complexes between the last quarter of the 3rd and the very beginning of the 2nd millennium BC cannot be excluded. Two groups of dates make it possible to confirm the stages in the development of the Garino culture.

At the early stage of the Garino culture, large vessels (up to 70%) with a restricted or straight mouth and a rounded bottom are widespread. There are rims with a bulge on the inside of the vessel. The Г-shaped rim form appears. The proportion of ornamentation with a toothed stamp reaches 60%. Sometimes the decoration is located on the inside of the rim. There is a horizontal row of round pit-like impressions under the rim on early Garino vessels. The surface of the vessels is densely filled with decoration patterns. The ornamentation is dominated by a long comb stamp. The motif of the "stepping comb" is widely represented. The presence of the Ayatskaya culture of ceramics of Trans-Ural origin is specifically recorded at the early stage.

Closed forms of vessels disappear at the later stage, and new forms of vessels appear: flat-bottomed with an open orifice and a bent rim. Vessels with wavy rims and raised cordons (Zayurchim I, Kamskiy Bor II, Rychino III) spread towards the end of the Garino era. The appearance of ceramics with raised wavy cordons and homogeneous ornamentation with a fine-toothed stamp in the late Garino complexes is interconnected with the emerging Seima-Turbino culture community of the region. This phenomenon is reflected in the typological proximity of ceramics with raised wavy cordons of the Kama region with the West Siberian Krotovo wares. The appearance of horizon pottery with raised cordons in the area of distribution of Garino pottery should be considered, along with the Chirkovo complexes of the Mari Volga region, in the framework of the formation of the Seima-Turbino phenomenon in the taiga zone of Eastern Europe (Solovyev 2000: 98; Denisov et al. 2012). There is a depletion of ornamentation. The motives of the "stepping comb" and the oblique lattice disappear in the late Garino era. The ornament becomes sparse, while some vessels have no ornamentation at all. It was during this period that the action of its own centre of metallurgy actively manifested. This period is also characterized by findings of metal weapons (spears - Zayurchim I, Sauz I) and ornaments of the Abashevo culture.

Obviously, contacts between the Bor and Garino populations began in the last quarter of the 3rd millennium BC, as evidenced by the materials of the Bor V settlement. It is likely that the Garino population assimilated the members of the Bor culture type at the beginning of the 2nd millennium BC. The end of the Garino culture is chronologically set in the middle of the 2nd millennium BC according to the finds of the Abashevo culture within the dwellings (Starushka, Krasnoe Plotbishche, etc.), supported by radiocarbon dates.

Acknowledgements

This study was supported by RFFR, project 1-20-49-590002 «Complex studies of Post-Neolithic and Chalkolithic cultures in the territory of the Middle Cis-Urals».

References

Bader 1954 = Бадер, О. Н. 1954. Стоянка Бор II и предананьинское время в Прикамье. *Советская Археология* XX (20): 180-212.

Bader 1961a = Бадер, О. Н. 1961a. *Поселения турбинского типа в Среднем Прикамье*. Материалы и Исследования по Археологии 99.

Bader 1961b = Бадер, О. Н. 1961b. Поселения у Бойцова и вопрос периодизации среднекамской бронзы. *Отчеты Камской (Воткинской) Археологической Экспедиции*. Вып. 2: 110-272.

Bader 1963 = Бадер, О. Н. 1963. *Древнейшая история Прикамья*. Москва.

Bader 1964 = Бадер, О. Н.*Древнейшие металлурги Приуралья*. Москва: Наука.

Bader & Kokarev 1959 = Бадер, О. Н & Кокарев, А. В. 1959. Поселение у Малого Борового озера. *Отчеты Камской (Воткинской) Археологической Экспедиции*. Вып. 2: 130-151.

Batueva & Lychagina 2018 = Батуева, Н. С. & Лычагина, Е. Л. 2018. Результаты исследования нео-энеолитической керамики с гребенчатым орнаментом Чашкинского микрорегиона. *Вестник Пермского университета* 1: 31-43.

Chairkina 2005 = Чаиркина, Н. М. 2005. *Энеолит Среднего Зауралья*. Екатеринбург.

Chernykh 1970 = Черных, Е. Н. 1970.*Древнейшая металлургия Урала и Поволжья*. Москва: Наука.

Chernykh & Kuzminykh 1989 = Черных, Е. Н. & Кузьминых С. В. 1989. *Древняя металлургия Северной Евразии (сейминско-турбинский феномен)*. Москва: Наука.

Denisov et al. 2012 = Денисов, В. П., Мельничук, А. Ф. & Тресков, С. А. 2012. Поселение Рычино III в Удмуртском Прикамье и валиковая керамика в позднегаринском культурном пространстве. *Труды Камской археолого-этнографической экспедиции* 8: 115 – 120.

Denisov & Melnichuk 2014 = Денисов, В. П. & Мельничук, А. Ф. 2014. Поселение Гагарское III в системе новоильиских древностей Пермского Приуралья. *Вестник Пермского университета* 1: 44-59.

Koksharov 2012 = Кокшаров, С. В. 2012. Первый металл Конды. *Вестник археологии, антропологии и этнографии* 4: 27-42.

Kuzminykh et al. 2013 = Кузьминых, С. В., Дегтярева, А. Д. & Денисов, В. П. 2013. Металлообработка гаринской культуры Верхнего и Среднего Прикамья (по данным аналитического исследования). *Вестник археологии, антропологии и этнографии* 4: 13-21.

Lychagina 2011 = Лычагина, Е. Л. 2011. Проблемы хронологии неолита – раннего энеолита Прикамья. *Вестник Пермского Университета* 1: 17-21.

Lychagina & Vybornov 2009 = Лычагина, Е. Л. & Выборнов, А. А. 2009. К вопросу о происхождении и хронологии новоильинской энеолитической культуры. *Научный Татарстан. Гуманитарные науки. Археология и история* 2: 33-37.

Meadows, J. 2020. Interpreting 14C ages of the total organic carbon content of prehistoric pottery. In *Радиоуглерод в археологии и палеоэкологии: прошлое, настоящее, будущее*: 54-57. Санкт-Петербург: ИИМК РАН.

Melnichuk 1990 = Мельничук, А. Ф. 1990. О памятниках борского типа в Прикамье. In *Энеолит лесного Урала и Поволжья*: 97-104. Ижевск: УИИЯЛ УрО АН СССР.

Melnichuk 2009 = Мельничук, А. Ф. 2009. Зауральские керамические комплексы на памятниках неолита и палеометалла Среднего Приуралья. In *Этнические взаимодействия на Южном Урале*: 14-17. Челябинск: ЮГУ.

Melnichuk 2013 = Мельничук, А. Ф. 2013. Хронология гаринской культуры в Среднем Прикамье. In *Периодизация и хронология в археологии эпохи раннего металла Восточной Европы*: 157-160. Санкт-Петербург: СКИФИЯ-принт.

Melnichuk et al. 2006 = Мельничук, А. Ф., Скорнякова, С. В. & Чурилов, Э. В. 2006. Стоянка Усть-Залазнушка II – новый памятник хуторского типа в камском неолите. *Вопросы археологии Поволжья* 4: 120-125.

Nagovitsin 1990 = Наговицин, Л. А. 1990. Культурно-хронологическое соотношение гаринских и борских памятников Прикамья. In *Энеолит лесного Урала и Поволжья*: 82–96. Ижевск: УИИЯЛ УрО АН СССР.

Prokoshev 1935 = Прокошев, Н. А. 1935. Работы на строительстве Пермской гидроэлектростанции (Средволгстрой). Район реки Чусовой. *Археологические работы Академии на новостройках в 1932-33 гг.* Т.1: 176-187. Москва – Ленинград.

Ryndina & Degtyareva 2002 = Рындина, Н.В. & Дегтярева, А.Д. 2002. *Энеолит и бронзовый* век. М: МГУ.

Shmidt 1935 = Шмидт, А. В. 1935. Работы на строительстве Пермской гидроэлектростанции. Введение. Район р. Камы. *Археологические работы Академии на новостройках в 1932-33 гг.* Т.1: 66-175. Москва – Ленинград.

Solovyev 2000 = Соловьев, Б. С. 2000. *Бронзовый век Марийского Поволжья*. Йошкар-Ола: МарНИИ.

Stavitskiy 2011 = Ставицкий, В. В. 2011. Проблема происхождения гаринской культуры. *Тверской археологический сборник*. Вып.8 Т.I: 229-231. Тверь: ТГОМ.

Vybornov 2008 = Выборнов, А. А. 2008. *Неолит Волго-Камья*. Самара.

Vybornov et al. 1984 = Выборнов, А. А., Обыденнов, М. Ф. & Обыденнова, Г. Т. 1984. Поселение Сауз I в устье реки Белая. In *Эпоха меди юга Восточной Европы*: 3-21. Куйбышев.

Vybornov et al. 2019 = Выборнов, А. А., Лычагина, Е. Л., Васильева, И. Н., Мельничук, А. Ф. & Кулькова, М. А. 2019. Новые данные о периодизации и хронологии новоильинских, гаринских и борских памятников Прикамья. *Вестник Пермского университета* 1: 34-47.

Kharlovka 1-6 on the Kola Peninsula: One of the Oldest Gressbakken House Sites in Northern Fennoscandia

Evgeniy M. Kolpakov, Anton I. Murashkin, Alevtina M. Kiseleva, Vladimir Ya. Shumkin & Kristiina Mannermaa

Evgeniy M. Kolpakov, Institute for the History of Material Culture, Russian Academy of Sciences, Department of Paleolithic; Dvortsovaya emb., 18, 191186 Saint-Petersburg, Russia, e-mail: eugenkolp@yandex.ru
Anton I. Murashkin, Institute for the History of Material Culture, Russian Academy of Sciences, Laboratory of cameral works, digital processing, documentation and keeping of archaeological collection; Dvortsovaya emb., 18, 191186 Saint-Petersburg, Russia, e-mail: aimurash@yandex.ru
Alevtina M. Kiseleva, Institute for the History of Material Culture, Russian Academy of Sciences, Department of Paleolithic; Dvortsovaya emb., 18, 191186 Saint-Petersburg, Russia, e-mail: aliakiseleva@mail.ru
Vladimir Ya. Shumkin, Institute for the History of Material Culture, Russian Academy of Sciences, Department of Paleolithic; Dvortsovaya emb., 18, 191186 Saint-Petersburg, Russia, e-mail: shumkinv@yandex.ru
Kristiina Mannermaa, Department of Cultures, archaeology, PO Box 59, 00014 University of Helsinki, Finland, e-mail: kristiina.mannermaa@helsinki.fi

Abstract

The remains of a semi-subterranean house were discovered at the Kharlovka 1-6 site on the Barents Sea coast of the Kola Peninsula, Russia. Despite suffering from erosion, features such as a double stone-boarded hearth and chimney pipe together with diagnostic finds allow us to attribute the house to the so-called Gressbakken type. The artefact assemblage includes stone tools, bone and antler tools and adornments, asbestos-tempered pottery, an amber pendant and a copper tubular bead. Numerous faunal remains are primarily represented by marine species of mammals and birds. Radiocarbon dates on charcoal point to the period 2600–2300 calBC, which is also supported by the artefact types, elevation of the cultural layer and construction type of the house. When accounting for the early dates of Kharlovka 1-6 and several other dwellings, it can be assumed that the Gressbakken house tradition emerged around 2600–2300 calBC. This date is somewhat earlier than previously thought.

1 Introduction

In 2017–2019, the Kola Archaeological Expedition of the Institute for the History of Material Culture of RAS undertook excavations at the Kharlovka 1-6 site on the Barents Sea coast of the Kola Peninsula, approximately 170 km east of the city of Murmansk (Fig. 1). The excavations have revealed the remains of a semi-subterranean house located on top of a sandy hillock 1 km northwest of the mouth of the Kharlovka River (Fig. 2–3). Some of the construction features and diagnostic finds allow us to attribute the excavated house to the so-called Gressbakken type.

Despite extensive destruction, the dwelling is rather exceptional for the Kola Peninsula region. The well-preserved organic material has yielded bone and antler tools and a large number of faunal remains, which provides us with data concerning subsistence strategies and seasonality.

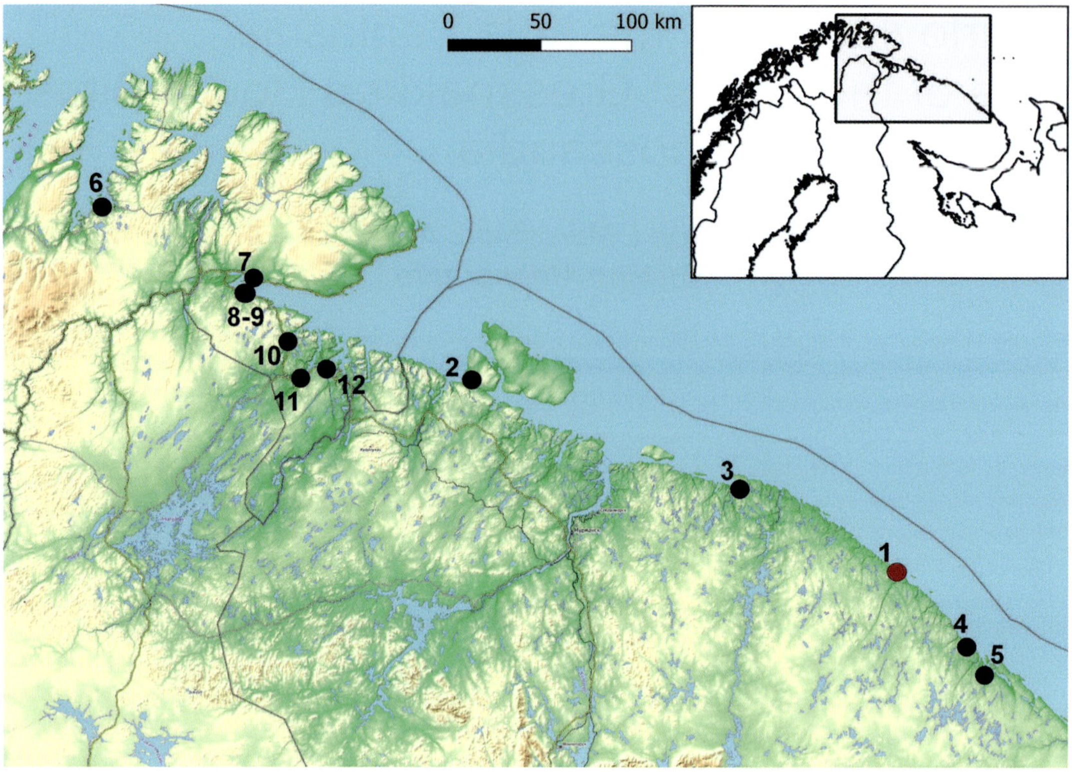

Figure 1. Overview map showing the location of Kharlovka 1-6 and other sites mentioned in the text. 1 – Kharlovka 1-6, 2 – Grottug (Rotojoki), 3 – Zavalishina 5, 4 – Dvorovaya, 5 – Ust-Drozdovka 3, 6 – Leirpollen, 7 – Bergeby, 8 – Gressbakken Nedre Vest, 9 – Advik, 10 – Bugøyfjord, 11 – Kalkillebukta, 12 – Høybukt. Drawing: A. Kiseleva.

The archaeological material from the site includes several unique finds made of copper and amber. These exotic goods reflect distant contacts that the ancient population of the Kola Peninsula had with southern regions. Together with one bone artefact manufactured with a metal implement, the copper item demonstrates one of the earliest occurrences of metal use in Arctic Europe. Finally, the radiocarbon dates point to the middle of the 3rd millennium calBC, making the excavated dwelling one of the oldest examples of the Gressbakken house tradition in northern Fennoscandia. A presence of similar constructive features of Gressbakken dwellings and dwellings of the taiga zone of Eastern Europe may also indicate contacts with southern and eastern territories.

2 Research history

To date, the Kola Archaeological Expedition has recorded 1,300 semi-subterranean and above-ground houses in the Murmansk region, of which a significant number date back to the prehistoric period. The largest clusters of houses have been recorded in Nokuev and Teriberka bays, along the Kildin Strait and on the Fisher and the Sredniy peninsulas. In total, 28 dwellings from the Neolithic – Bronze Age (Early Metal Age) have been excavated (Kolpakov et al. 2020: 279–288, 296, Fig. 18), including a house at Kharlovka 1-6 that was studied in recent years.

The mouth of the Kharlovka River was first explored by Nina Gurina in 1947, but she discovered the Kharlovka 1-6 site in

Figure 2. The topographic plan of the Kharlovka 1-6 site. Drawing: E. Kolpakov.

Figure 3. The location of the Kharlovka 1-6 site. Aerial view from the south. Photo: E. Kolpakov.

1974. The researcher documented six concentrations of lithic artefacts on an exposed sandy surface (Fig. 2) and excavated 40 square metres near concentration no. 2. A hearth and numerous slate and quartz flakes were registered in the excavated area (Gurina 1974). Later, Vladimir Shumkin again surveyed the site in 1978. In addition to stone artefacts and one ceramic fragment, he found a large number of marine mammal bones on the sandy surface of the hillock. However, he assumed that these remains were modern as opposed to prehistoric (Shumkin 1978).

In 2014, we surveyed the Kharlovka River area and the sites that had been investigated earlier, including the sandy hillock with faunal remains. We documented a bone- and artefact-rich cultural layer on top of the slope, from which the date 2940–2570 calBC (SPb-1561) was derived (see below in the 6. Dating section). Though the date was derived from seal bones and the marine reservoir effect had a certain impact, the evidence showed that the layer likely dated back to the Stone Age. Topographic features of the area also indicate the prehistoric character of the site (Kolpakov 2014).

Archaeological sites with good preservation of organic matter are extremely rare in the northern part of Fennoscandia. Previously, only three settlements (Ust-Drozdovka 3, Mayak 2, Zavalishina 5) with large amounts of faunal remains and bone tools have been excavated on the Kola Peninsula. Therefore, Kharlovka 1-6 is particularly interesting to researchers.

3 Excavation results

During three field seasons, 44 square metres were studied. We excavated a turf-covered section of a hillock not destroyed by erosion.

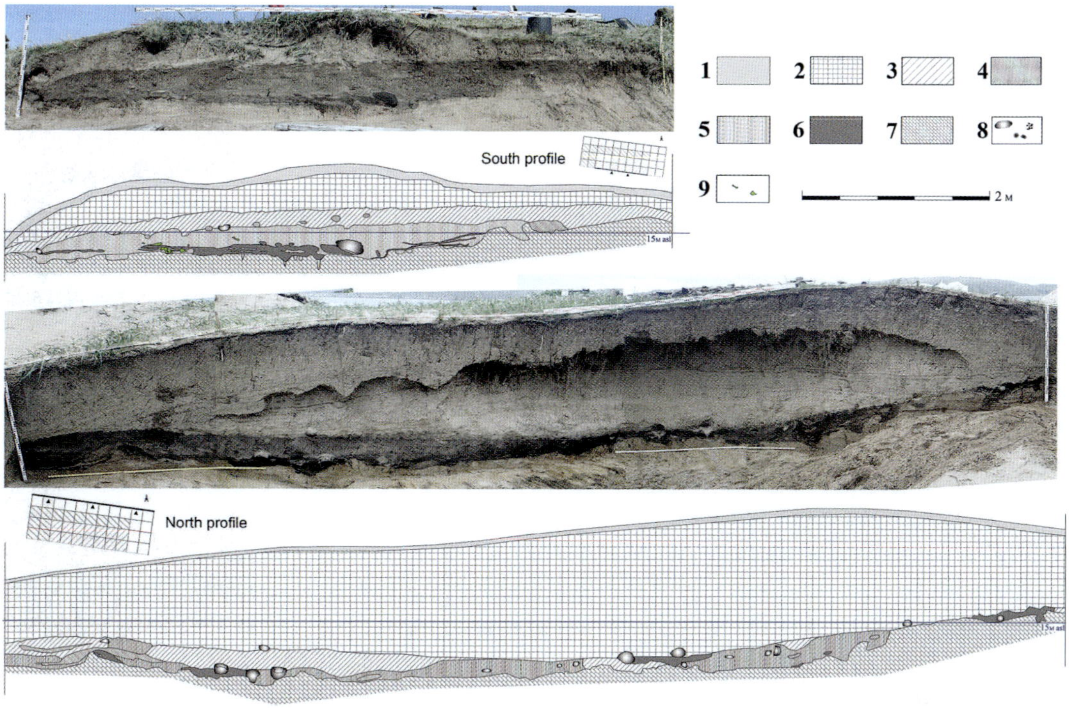

Figure 4. Southern and northern profiles of the excavation area at Kharlovka 1-6. 1 – turf, 2 – light grey sand, 3 – dark brown organic reach sand with lenses (a cultural layer), 4 – grey sand, 5 – grey sand with charcoal and bone fragments (a cultural layer), 6 – dark organic reach sand, 7 – light grey sand (natural layer), 8 – stones, 9 – bones. Drawing: E. Kolpakov.

In addition, archaeological and faunal material was collected from the destroyed surface to the south and southeast of the excavation area. During the excavation process, it became clear that the excavation area contains the remains of a badly damaged house. The southern part of the house and the upper part of its walls have been destroyed by erosion. The layer of the house lies at an elevation of 14,3–15,3 m in the Baltic elevation system.

3.1 Stratigraphy

The stratigraphy of the excavated area consisted of a thin turf layer (3–6 cm), a sand layer (up to 140 cm thick) and a cultural layer underlain by sand (Fig. 4). Both layers of sand – those covering and underlying the cultural layer – were light grey, contained almost no archaeological finds and appeared to be dune deposits. The occupation layer was grey and brown sand containing bone refuse, charcoal, fire-cracked stones, pebbles, artefacts and occasional shell fragments. The cultural layer, reaching a depth of 55 cm, was thickest at the centre and became thinner towards the walls.

The upper part of the cultural layer was mostly brown, contained many fire-cracked stones and stone artefacts in particular, and included several dark or light grey sand lenses. This was followed by a layer of dark grey sand with a large amount of charcoal. The layer under the hearth and stone accumulations was black. It is possible that the brown and grey sand layers indicate different stages in the existence of the house pit, but distinguishing them more precisely is difficult due to the later disturbance and irregularity in their deposition. Moreover, the archaeological material from the layers does not show

Figure 5. The plan view from the North (above) and profile view from the South (below) of the double hearth recovered at Kharlovka 1-6. Photo: E. Kolpakov.

any chronological discrepancy. Therefore, we tend to consider them as one complex.

The profile of the house proved it was a semi-subterranean dwelling. The cultural layer on the eastern and western sides of the excavation area was considerably thinner than in the centre or else it was almost completely absent, which most likely indicates the borders of the dwelling chamber. It is important that the distribution of the artefacts directly correlates with the thinning of the cultural layer. The cultural layer becomes thicker again in the eastern part of the dwelling, which can indicate a midden structure or the chamber of an adjacent dwelling.

3.2 The house structure

Our excavation has revealed some structural features of the dwelling. We identified the remains of a double hearth close to the destroyed area in the centre of the house chamber: two lines of flattened stones up to 45 × 20 cm in size raised on their sides (Fig. 5). Such double hearths are quite specific and are registered in most Gressbakken dwellings. Therefore, it is possible to determine other parts of the structure. The hearth was placed along the longitudinal axis of the house. Its length reached 3,5 metres, including the space between the two parts. Only one row of stones consisting of two parts remained. This row represents the northern half of the hearth. The southern half of the hearth was likely destroyed by erosion and fell down the slope.

An oval concentration (100 × 60 cm) of pebbles extended perpendicular to the hearth to the north of it, which included long animal bones situated perpendicular to the concentration's axis. Based on analogies from other excavated houses, we assume that this feature represents a chimney pipe running from the hearth to the back wall. We suggest that similar chimney pipes have been recorded at Zavalishina 5 (houses 6 and 7) (Kolpakov et al. 2016: Fig. 8), Ust-Drozdovka 3 (house 5) (Helskog et al. in press), most likely at Grottug (Seitsonen 2006: Fig. 3) on the Kola Peninsula and at Kalkillebukta (house 7) in Varangerfjord (Schanche 1994: Fig. 14). A. Niemi and J.

Oppvang reported the same construction in house 11 at Nyelv Nedre Vest and interpreted it to be an air channel (Niemi & Oppvang 2018: 50, Fig. 42).

Unfortunately, the northern wall of the house has not been found. Furthermore, there is a 20–35-cm downward drop in the cultural layer in comparison with the centre of the chamber and the level of the hearth. Such a feature is not found in other Gressbakken houses. The absence of the northern wall of the chamber can most likely be explained by post-deposited subsidence of the soil. On the other hand, another object like a large pit or semi-subterranean dwelling could have later disturbed or cut into the excavated house chamber, but no distinct evidence of such a process has been recorded. Otherwise, if we assume the chamber spreads in this direction, the house would have been extremely large and would have no analogies.

Four concentrations of dark sand, charcoal, pebbles and fire-cracked stones are likely to be found outside the house and relate to some other structure. Stones in the larger clusters under the northern wall of the excavation varied considerably in size (up to 37 cm) and had been laid out very densely. Decomposed wood fragments and animal bones were also found there.

3.3 Interpretation

Comparing the dwelling site with other houses allows us to offer an approximate "theoretical" reconstruction of the revealed structure (Fig. 6). The dimensions of the hearth and the distance between the eastern and western walls of the chamber are used to determine the measures of the dwelling. The dimensions of the dwelling chamber may well have been approximately 10 by 4 metres. It

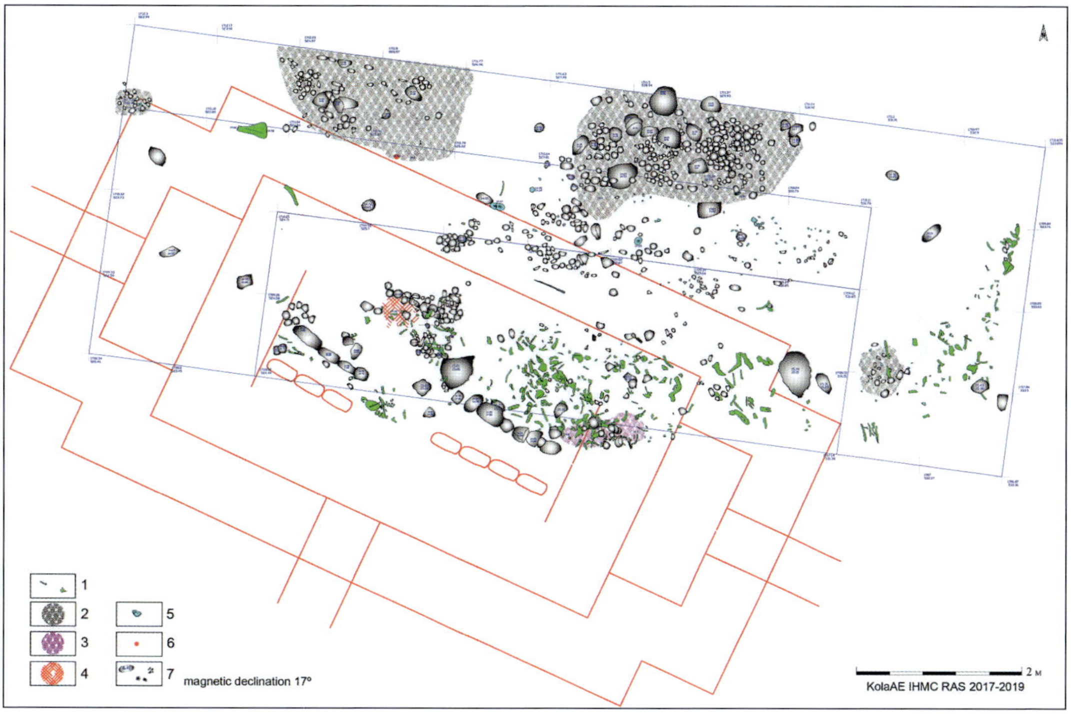

Figure 6. Plan of the excavated area at Kharlovka 1-6 and a supposed reconstruction of the house (red line). 1 – bones, 2 – charcoal-rich spots, 3 – shell-fragment-rich spots, 4 – burned soil, 5 – large artefacts, 6 – ochre, 7 – stones. Drawing: E. Kolpakov.

is impossible to estimate the precise depth of the house, though, as its uppermost portion was eroded. However, the remaining part of the structure was as deep as 60–70 cm. The double stone-bordered hearth and the chimney pipe were placed in the middle of the chamber. It seems that a raised platform had been constructed in the eastern part of the chamber. The southern and northern walls of the chamber were absent due to erosion or post-deposited disturbances.

There is also a possibility that the abandoned house was used as a deposit for animal remains and stone refuse. Many animal bones, quartz flakes and shell fragments were concentrated in the southern part of the dwelling above the hearth. It is notable that animal remains in the concentration were occasionally found in anatomical order.

3.4 Do the archaeological remains uncovered at Kharlovka 1-6 constitute a dwelling?

This question can most likely be answered in the affirmative, and for good reason. A double hearth or two hearths are represented in the dwelling site. Even if we dispute the existence of a second row of stones forming a hearth fence in the past, an undeniable black layer in the place of the hearth remains. Therefore, the hearth certainly exists. The eastern and western walls of the house pit, up to half a metre or more deep, have been uncovered and are beyond dispute. The southern wall, along with the entire southern half of the dwelling, has been completely destroyed. Therefore, the absence of the southern wall of the dwelling in the excavation is logical.

The problem is the absence of the proposed dwelling's northern wall in the excavation. We compared the dimensions of the structure with other excavated dwellings of different types and concluded that the northern wall should have been found within the limits of our excavation. However, no such wall has been found. It can be assumed that the dwelling was oriented with the long axis in a north-south direction instead of an east-west one. However, in that case the double hearth would have been located across the dwelling, which is hardly possible, and such has not been found in previously studied dwellings.

The lowering of the dwelling floor via the ledge below the level of the hearth also represents a problem when interpreting the excavated site. Dwellings with such a constructive element are entirely unknown at other similar sites. Could the ledge have formed as a result of a post-deposition subsidence? Such a possibility cannot be ruled out, given the clear destruction and deformations observed at the site.

Could the ledge have formed as a result of one dwelling pit cutting into another pit? This option is extremely unlikely, since no traces of this activity have been found in the cultural layer.

Thus, we can conclude that while the remains of a deepened dwelling were uncovered in the excavation at Kharlovka 1-6, many of its structural elements remain unclear.

3.5 Is this a dwelling of the Gressbakken type?

It must be borne in mind that the type itself does not have an unequivocal definition.

Povl Simonsen first described Gressbakken dwellings in 1961 (Simonsen 1961: 510). His description included a combination of features such as a large rectangular or oval semi-subterranean chamber, a stone-framed double hearth placed along the longitudinal axis, raised platforms in each end of the chamber and multiple "entrance passages." The description of 1979 omitted one important feature, namely multiple "entrance passages" (Hood 2017: 14). In the following years, excavations of Gressbakken dwellings have continued (Helskog 1983; Hesjedal et al. 1996; Hesjedal et al. 2009; Schanche 1994), as has the discussion of the problems associated with the "Gressbakken phase" (Olsen 1994: 71–76; Hesjedal et al. 1996: 211–219; Hood 2017: 12–16, 26–28; Niemi

& Oppvang 2018: 75–78). But the concept of the "Gressbakken dwelling" was eroded. Dwellings with only one, two or three of the above-mentioned features were classified as this type (for example dwelling 21 from Gressbakken NØ (deep chamber and one hearth) (Schanche 1994: 54) and dwellings 82 and 83 from Slettnes 5C (large but shallow chamber, double (?) hearth, one entrance)) (Hesjedal et al. 1996: 123–126, 136, 214–218). In some cases, a reason for the classification was artefact types and radiocarbon dates. For example, Schanche considered dwellings 17–20 from Iversfjord to be Gressbakken dwellings (Schanche 1994: 60–62), though E. Helskog did not (Helskog 1983: 86). As a result, the "Gressbakken dwelling" type is used by most researchers of northern Fennoscandia, but each can use discrete features from a long list to define the type.

It is more realistic to discuss the similarity of certain structural elements and their combination. As a whole, this combination includes a large rectangular-oval, semi-subterranean chamber, a stone-framed double hearth placed along the longitudinal axis, raised platforms in each end of the chamber and multiple "entrance passages."

A double hearth (according to our reconstruction) is present in the excavation at Kharlovka 1-6. The dwelling also includes a deep chamber. There is a rise in the floor in the eastern part of the chamber, which is similar to the platforms at the ends of the chamber often found in other dwellings of the Gressbakken culture. Stones in large clusters to the north of the hearth are usual in Gressbakken houses as well. In fact, these are the only structural elements that can be associated with dwellings of the Gressbakken type in the excavated dwelling.

Thus, there are two reasons why the dwelling excavated at Kharlovka 1-6 cannot be unquestionably attributed to the Gressbakken type. First, the vagueness of the Gressbakken type itself. Second, the lack of structural elements and flaws in existing ones of the dwelling discovered in the excavations. At the same time, the types of collected artefacts correspond well with those associated with the last phase of the Young Stone Age and Early Metal Age (end of the 3rd – 2nd millennium calBC) and frequently found in Gressbakken dwellings (see below).

Therefore, we can conclude that the assemblage excavated at Kharlovka 1-6 is 1) a semi-subterranean dwelling, 2) belongs to the Gressbakken culture/phase and 3) has a number of structural features common to dwellings of this culture.

4 Artefact assemblage

The artefact assemblage from the Kharlovka 1-6 house includes a large quantity of lithic material (43 807 pcs.), a number of bone tools (48 pcs.), ceramic sherds (32 pcs.), an amber pendant and a copper tubular bead. A number of unmodified artefacts exist as well: quartz (2 pcs.), sandstone (1 pc.) and pumice (9 pcs.) pebbles, pieces of ochre (5 pcs.) and asbestos (16 pcs.). It also contains the bones of mammals, fish and birds. The majority of the material was found within the preserved house layer, but a fair quantity was also collected from the sandy surface next to the excavated area.

4.1 Lithics

Table 1 describes the general characteristics of the lithic assemblage. Local sources of raw materials were used for the manufacturing of quartz tools (pebbles and pieces of vein quartz). The knapping technique was aimed at obtaining flakes. Slate is also widely represented at the site. However, the small size of the flakes, the high percentage of debris from ground tools and a case of their refurbishing into bipolar cores might indicate a scarcity of this material. The number of chert and sandstone flakes in the assemblage is negligible.

The assemblage is dominated by quartz tools and debitage. Various types of scrapers are present: single-edged scrapers and double-edged scrapers with convex and concave edges (Fig. 7: 1–3). Knives (Fig. 7: 5), piercing tools (Fig. 7: 4) and a combination of

Techno-logical group	Type	Lithic material							Total
		quartz	slate	chert	soap-stone	sand-stone	granit	pum-ice	
Debitage	Precore, Core	397	2						399
	Chip	42							42
	Flake, blade (including biface trimming flakes)	10850	563	15		11			11439
	Micro-debitage	30655	354	4					31013
	Sawed-out fragment		4						4
Tools and preforms	Retouched flake, blade	206	7	2					215
	Scraper	390	1	1					392
	Piercing tool, point	21							21
	Knife	29							29
	Composite tool	4							4
	Retouched arrowhead	4							4
	Bifacial arrowhead, preforms	1	7						8
	Bifacial preform	1	5						6
	Slate plate preform		3						3
	Polished knife		14						14
	Polished arrowhead		23						23
	Polished rod		1						1
	Wood-chopping tool, preform		4						4
	Polished tool fragment		4						4
	Flake detached from a ground tool		114						114
	Plate with traces of grinding		2		1	1			4
Abrasive tools	Grinding slab		4			4			8
	Saw					2			2
	Grinding slab with a groove							5	5
Other	Hammer stone	1					12		13
	Anvil						1		1
	Notched club						1		1
	Weight-stone						1		1
Total		42601	1112	22	1	18	15	5	43774

Table 1. Kharlovka 1-6. Lithic artefact assemblage.

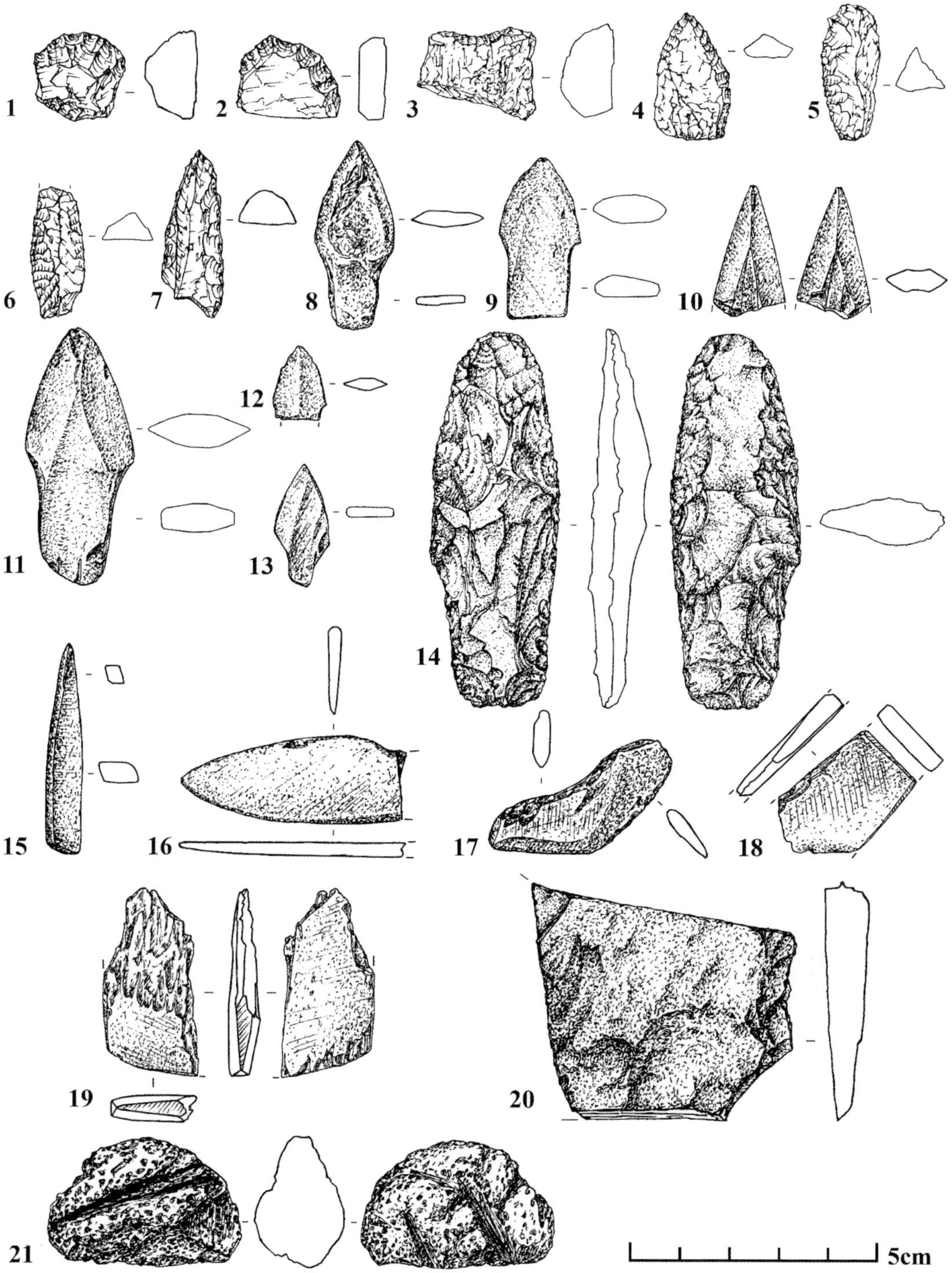

Figure 7. Lithic artefacts from the Kharlovka 1-6 site. 1–3 – scrapers, 4 – piercing tool, 5 – knife, 6–7 – retouched arrowheads, 8–13 – polished arrowheads, 14 – bifacial preform, 15 – polished rod, 16–18 – polished single-edged knives, 19 – plate with traces of grinding, 20 – saw, 21 – grinding slab with grooves. 1–7 – quartz, 8–18 – slate, 19 – soapstone, 20 – sandstone, 21 – pumice. Drawing: A. Malutina.

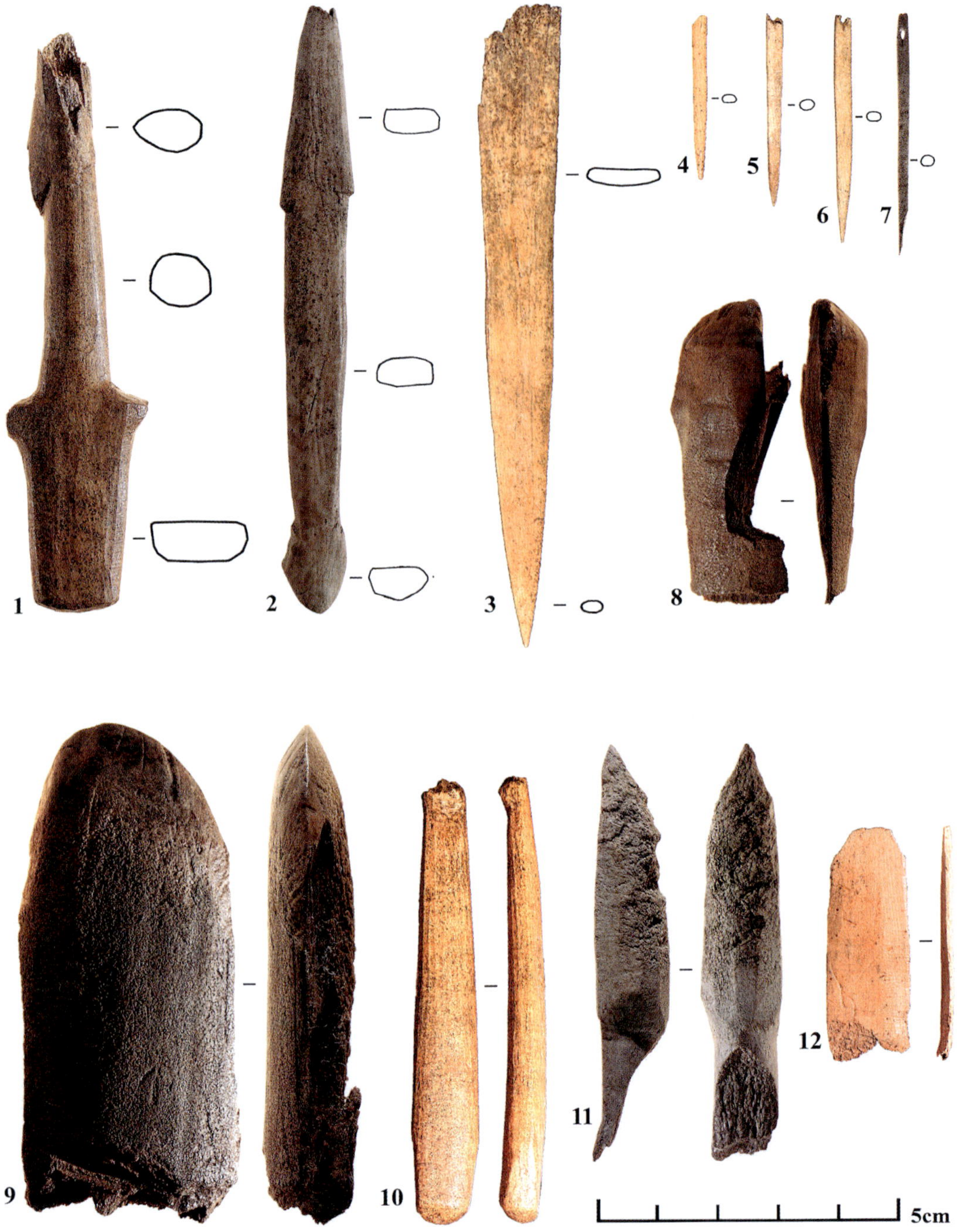

Figure 8. Bone and antler artefacts from the Kharlovka 1-6 site. 1 – barbed harpoon head, 2 – leister head, 3 – piercing tool, 4–7 – needles, 8–9 – chisels, 10 – crochet hook, 11 – preform for an arrowhead, 12 – fragment with traces of processing with metal tool. Photo: D. Shekhirev, A. Malutina.

tools (piercing tool–scraper) are also present in the assemblage. The quartz material also includes one blade and 205 retouched flakes. Most of them have an irregular retouch, which does not form a clear working edge and might be a result of usage. Other items with 2–3 retouch scars might obviously constitute tool fragments.

Four quartz tools are considered to be fragments of arrowheads. They have two parallel edges on the dorsal surface with an abrupt or semi-abrupt retouch (Fig. 7: 6); one of them has retained its sharp point (Fig. 7: 7). In addition, there is a base fragment of a bifacial arrowhead and a preform of a bifacial tool. Two implements classified as piercing tools and characterised by a regular semi-abrupt retouch at two edges that converge at an acute angle could also be fragments of quartz arrowheads.

The next largest component of the lithic assemblage is slate. Twenty-three ground slate arrowheads and their fragments exist, 11 of which are diagnostically complete (all tanged) (Fig. 7: 8, 9, 11–13), eight of which are tang fragments and four of which are point fragments. One of the fragments has longitudinal grooves on both surfaces, which is one of the main characteristics of the Sunderøy type (Fig. 7: 10), widespread throughout northern Fennoscandia during the Late Neolithic and Bronze Age (Gjessing 1942, p. 172–174; Rankama 1986). The assemblage contains six fragments (2 distal, 4 medial) and one complete preform of bifacial arrowheads. All of them have been treated with irregular retouching and have an asymmetrical cross section, while one preform has a tang (Fig. 7: 14). Most likely they are preforms for ground arrowheads.

Polished knives (14 pcs.) are the second largest group of slate tools. Only one intact knife exists in the assemblage (Fig. 7: 17), with the rest represented by fragments. All of them are single-edged knives, with the angle of convergence of blade and handle at approximately 40–50 degrees (Fig. 7: 16, 18). As far as can be observed from the fragments, all the knives are miniature in size.

Wood-chopping tools are represented by one miniature chisel, one small preform and two fragments of adzes (butt and blade). The assemblage also includes 114 flakes detached from ground tools, some of which were undoubtedly knapped off from massive wood-chopping tools. The slate material also includes one polished rod with a pointed tip (Fig. 7: 15), four undefined fragments of polished tools and two plate fragments with traces of grinding. Four fragments of bifacial preforms were left in the initial stage of manufacturing.

The technological group of abrasive tools includes grinding slabs made of slate (4 pcs.) and sandstone (4 pcs.) as well as two fragments of sandstone saws (Fig. 7: 20). The use of sawing is also exemplified by four slate sawed-out fragments. There are five small rounded pumice pebbles with narrow grooves from sharpening objects with small cross sections (needles, awls?) (Fig. 7: 21). Other finds include a miniature plate made of soapstone grounded on the flat and lateral side surfaces (Fig. 7: 19), with hammer stones made of granite (12 pcs.) and quartz (1 pc.) pebbles. The material also contains a granite anvil, a sinker and a notched club. The club has a transverse ring groove for tying, and the faces of the club show traces of smashing.

4.2 Bone tools

The bone and antler assemblage are relatively abundant. A total of 48 implements and fragments of worked bones were found, with 19 tools being finished and archaeologically complete and the rest consisting of small fragments or preforms. The technological and use-wear analysis has revealed that the tools and preforms made of reindeer antler outnumber those made of reindeer bone (Murashkin et al. 2019: 92). This corresponds with the ratio of such raw materials found at Ust-Drozdovka 3 and Gressbakken Nedre Vest (Hodgetts & Rahemtulla 2001).

The presence of preforms and manufacturing debris apparently indicates that the

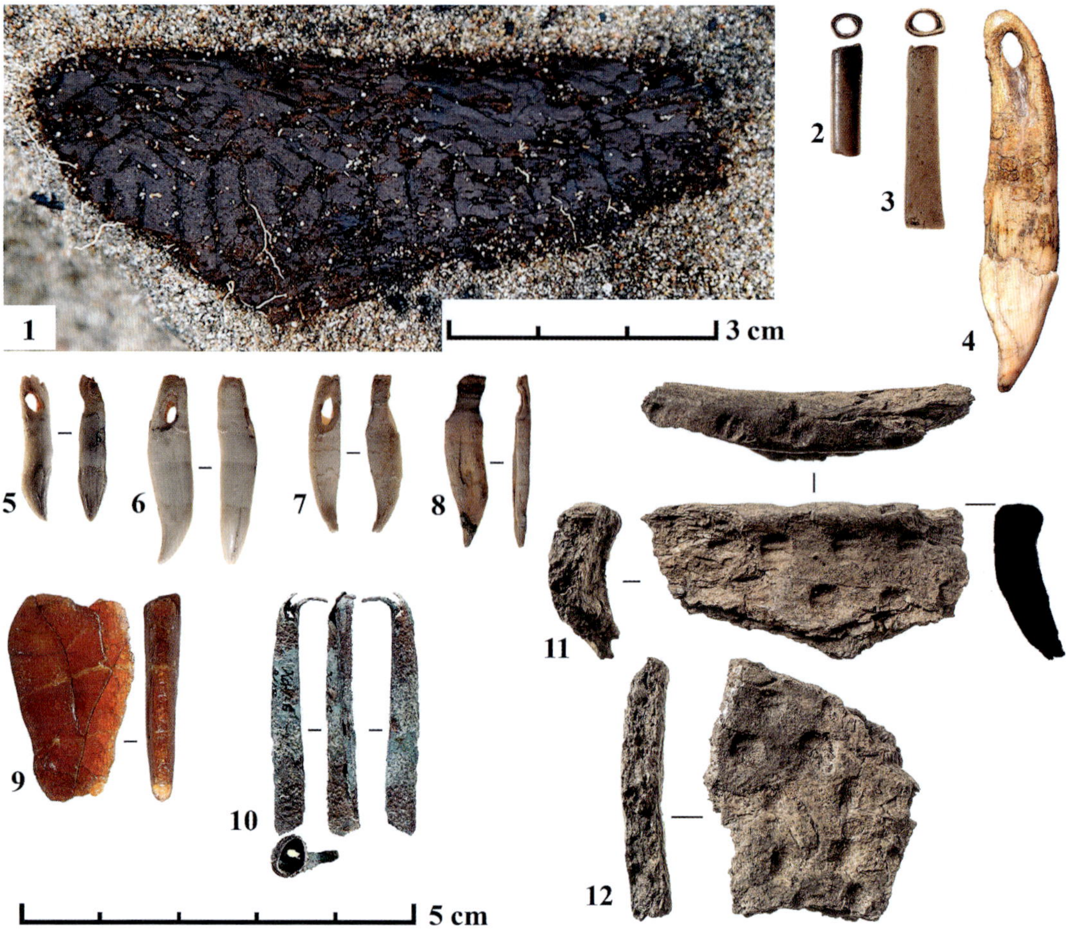

Figure 9. Artefacts from the Kharlovka 1-6 site. 1 – decorated bone plate, 2–3 – bone tubular spacer beads, 4 – pendant made of elk tooth, 5–8 – pendants made of seal teeth, 9 – amber pendant, 10 – copper tubular bead, 11–12 – ceramic fragments. Photo: E. Kolpakov, D. Shekhirev, Yu. Rumiantsev.

production of implements took place at the site. The preforms were predominantly produced using the method of longitudinal and transverse splitting and breaking along precut grooves or cracks. The implements were manufactured by planing, scraping and cutting. The final processing included grinding and polishing (Murashkin et al. 2019: 92).

The fishing and hunting equipment includes a barbed harpoon head (Fig. 8: 1), a leister (Fig. 8: 2) and a preform of an arrowhead (Fig. 8: 11). The leister is characterised by two small barbs and an expanded rounded base. The harpoon head has one barb and a rectangular base and, according to the typological system of Murashkin and Kiseleva (2018: 113–117), it belongs to a time-diagnostic type hG2. Harpoons of this type were widespread during the Gressbakken phase in northeastern Norway (Murashkin et al. 2019: 89–91).

Other tools include five needles (Fig. 8: 4–7), three chisels (Fig. 8: 8, 9), a crochet hook (Fig. 8: 10) and two awls. All the needles are characterised by a rectangular cross section and the presence of a cut and pierced eye. The surface of the needles bears polishing and linear traces associated with leather processing (Murashkin et al. 2019: 93–95). The traces on the rounded chisel edge also relate to leather processing (probably to fleshing) (Murashkin et al. 2019: 97–99). The crochet hook with a small projection on the stem has clear analogies with finds from houses at Gressbakken Nedre Vest (Simonsen 1961: Fig. 142: a) and Mayak 2 (Gurina 1997: Fig. 62: 23–26, 28).

Adornments comprise the most numerous category. They include five teeth pendants with holes made by cutting and piercing (four from seal canine teeth (Fig. 9: 5–8) and one from an elk's incisor tooth (Fig. 9: 4)). Three tubular spacer beads made of bone (probably bird bone) were also found (Fig. 9: 2, 3). An interesting find is a bone plate with geometric patterns. Unfortunately, it was in very bad condition (Fig. 9: 1). Artefacts with a similar pattern were found at Mayak 2 (Gurina 1997: Fig. 54: 3, 4, 14), Ust-Drozdovka 3 (Murashkin et al. 2019: Fig. 4: 37) and Gressbakken Nedre Vest (Simonsen 1961: Fig. 143: l).

Use-wear analysis showed that virtually all bone/antler implements were produced with stone tools. However, one small fragment of an artefact made of antler (?) bears traces of planing with a metal instrument (Fig. 8: 12), which is unique evidence from the period.

4.3 Ceramics

The assemblage includes 32 small pottery sherds, most of which do not have an outer or inner surface. The colour of the pottery is light brown or grey, and the thickness ranges from 4 to 9 mm. All the fragments are tempered with asbestos, while one of them also has mica as an admixture (Fig. 9: 11, 12).

Only one rim was found in the house (Fig. 9: 11). This sherd is 6 mm thick and slightly profiled. Decoration consists solely of short comb-like impressions running in horizontal rows on the outer surface and the rim edge. Similar vessels have been found at different sites in Karelia. They belong to pottery of the Palayguba type dated from 2500 to 1750 calBC (Nordqvist 2018: Fig. 19: e–h; Zhul'nikov 1999: 53–55, Fig. 37–39; 2005; Zhul'nikov in press). Vessels from Sumozero XV (Zhul'nikov 2005: Fig. 89: 2), Bohta II (Zhul'nikov 2005: Fig. 88: 8) and Sulgu III (Zhul'nikov 1999: Fig. 37: 4, 8) have an almost identical ornamentation and rim shape. On the Kola Peninsula, pottery of this type was also collected from the Mayak 2 site (Zhul'nikov 2005: 29–31; Zhul'nikov in press).

4.4 Exotic goods

Both the amber pendant and the copper tubular bead were derived from the eroded slope area. The amber adornment is characterised by an oblong drop-shape form and circular perforations on the broader side (Fig. 9: 9). On the Kola Peninsula, amber artefacts were found at Mys Semerka (Gurina 1997: Fig. 45: 7) and Mayak 2 (Gurina 1997: Fig. 62: 3). Both were found in cultural layers that contain finds from the Neolithic to the Iron Age, so their dating is difficult. Several amber artefacts are known to exist in northern Norway; these mostly date back to the 4th millennium calBC (Ramstad 2006). However, in Karelia the flow of amber greatly increases from the second half of the 3rd millennium calBC onwards (Zhul'nikov 2008). Overall, we believe the find from Kharlovka 1-6 is typologically unrepresentative and does not provide a reference for dating. The small tubular bead is 32 mm long and 5 mm in diameter, rolled up with a thin sheet of copper in one and a half turns (Fig. 9: 10) and with traces of bending and breaking at one end. XRF analysis has shown that the tubular bead is made of pure copper. This find is unique for the Kola Peninsula. A similar product was found at the Lillberget settlement in northern Sweden (Halén 1994: Fig. 183, 188). It was originally dated to 3900 calBC, but current analysis suggests that the settlement was inhabited for a long time and the narrow dating of the tubular bead is questioned (Nordqvist et al. 2011). Tubular beads and bracelets were also found in some burial grounds of the Fat'yanovo culture on the Upper Volga (Krainov 1987: Fig. 35: 4, 20–21), which dated back to the 3rd millennium calBC (Krenke 2019).

5 Faunal remains

The osteological analysis is preliminary. Animal bones were found in most of the investigated area. Their level of preservation varies greatly: from almost perfect condition to complete decomposition. Therefore, the number of unidentifiable bones is large. In total, 10,678 bones or their fragments were investigated, of which 3,953 have been identified at the level of species, genus or family. The analysed materials include all mammal remains and bird bones excavated during the years 2017 and 2018, while the rest of the bird and fish bones have not been properly analysed yet. Also, most of the small, burned bone fragments have not been counted and examined. Osteological analysis of the mammal bones was carried out by Ekaterina Petrova (Zoological Institute RAS, Saint Petersburg), while the bird bones were identified by Natalia Volkova (Paleontological Institute RAS, Moscow) and Kristiina Mannermaa (University of Helsinki). The distribution of animal bones by species is represented in Table 2.

The mammal assemblage is dominated by marine fauna, especially harp seal (*Phoca groenlandica*) bones, which comprised 96,4% of the identifiable mammal bones. The remains are represented by all parts of the skeleton, but carpals, tarsals and phalanges predominate in number. Some of the bones were burned. The examined remains belong to adults, sub-adults and juvenile seals. However, the bone remains of sub-adult and juvenile animals outnumber those of the adults. The presence of juveniles indicates that at least some of the harp seals were killed in the spring and early summer.

The next largest marine mammal species is walrus, consisting of 37 pieces. Most of the walrus bones belong to an adult animal, while one bone belongs to a juvenile. There are also cetaceans, represented by two fragmented bones.

Of the terrestrial mammals, wild reindeer and red fox (*Vulpes vulpes*)/arctic fox (*Vulpes lagopus*) are dominant, though a few Eurasian elk (*Alces alces*), mountain hare (*Lepus timidus*) and Eurasian beaver (*Castor fiber*) bones also occur. The bones of these species belong to adults.

Bird remains are represented by marine birds, shore birds, forest birds and predator birds. They belong to ten families, of which gulls (Laridae), ducks (Anatidae) and auks (Alcidae) are most numerous. Other aquatic bird species include cormorants (*Phalacrocorax* sp) and pomarine jaeger (*Stercorarius pomarinus*). Forest and predator birds consist of hazel grouse (*Bonasa bonasia*), ptarmigan (*Lagopus* sp), common curlew (*Numenius arquata*) and peregrine falcon (*Falco peregrinus*). Wing bones are the most common anatomic elements for all material, but significant differences exist between most common families (legs are more common in auks than in ducks and gulls). In the 2017 assemblage, approximately 20% of the bones were also burned.

Most of the identified species are migratory and live along the shoreline of the Kola Peninsula only during spring, summer and autumn. Three juvenile birds – two subads and one pullus – were identified in the material. Two of them represent the common guillemot (*Uria aalge*)/razorbill (*Alca torda*) and peregrine falcon. These species are late breeders and demonstrate a late summer hunting season. One humerus of a large gull most likely contains a medullary bone that is present inside bones only during breeding (during egg formation). The presence of the medullary bone and young birds indicate that these birds were caught during the summertime. Hazel grouse and unspecified ptarmigan are local birds in the material, and they could be hunted year round.

Species	NISP	MNI
MAMMALS (2017, 2018, 2019)		
Harp seal (*Pagophilus groenlandicus*)	3718	25
Walrus (*Odobenus rosmarus*)	37	2
Cataceans (*Cetacea indet.*)	2	1
Reindeer (*Rangifer tarandus*)	32	2
Elk (*Alces alces*)	2	1
Fox/arctic fox (*Vulpes sp.*)	44	2
White hare (*Lepus timidus*)	1	1
Hare (*Lepus sp.*)	8	1
European beaver (*Castor fiber*)	1	1
Red-backed mouse (*Clethrionomys sp.*)	1	1
Red-backed lemming (*Myopus schisticolor*)	3	1
Bank vole (*Arvicola amphibia*)	3	1
Field vole (*Microtus sp.*)	1	1
Indeterminate bones of small mammals	9	
Indeterminate bones of large mammals	417	
Indeterminate bones	6234	
BIRDS (2017, 2018)		
Mallard (*Anas platyrhynchos*)?	1	
Velvet scoter/ goosander (*Melanitta fusca/Mergus merganser*)	1	
Goosander/ red-breasted merganser (*Mergus merganser/Mergus serrator*)	1	
Common eider (*Somateria mollissima*)	6	
King eider (*Somateria spectabilis*)	7	
Eider (*Somateria sp.*)	1	
Ducks (*Anatidae*)	5	
Hazel grouse (*Bonasa bonasia*)	4	
Willow/rock ptarmigan (*Lagopus lagopus/L. muta*)	1	
Black-throated loon (*Gavia arctica*)	1	
Great cormorant (*Phalacrocorax carbo*)	2	
Peregrine falcon (*Falco peregrinus*)	1	
Eurasian oystercatcher (*Haematopus ostralegus*)?	1	
Common curlew (*Numenius arquata*)	2	
Pomarine skua (*Stercorarius pomarinus*)	5	
Common gull (*Larus canus*)?	1	
Lesser black-backed gull (*Larus fuscus*)	6	
European herring gull (*Larus argentatus*)	1	
Great black-backed gull (*Larus marinus*)	11	
European herring gull/ Lesser black-backed gull (*Larus argentatus/L. fuscus*)	10	
European herring gull/Great black-backed gull (*Larus argentatus/L. marinus*)	7	
Great black-backed gull/glaucous gull (*Larus marinus/L. hyperboreus*)	2	
Brünnich´s guillemot (*Uria lomvia*)	1	
Common guillemot/Brünnich´s guillemot (*Uria aalge/Uria lomvia*)	7	
Razorbill (*Alca torda*)	1	
Razorbill (*Alca torda*)?	1	
Common guillemot/Brünnich´s guillemot/ razorbill (*Uria aalge/U. Lomvia/Alca torda*)	11	
White wagtail (*Motacilla alba*)?	1	
Atlantic puffin (*Fratercula arctica*)	1	
Indeterminate bird bones	65	
Total	**10678**	

Table 2. Kharlovka 1-6. Faunal assemblage.

Preliminary analysis of the fish bone assemblage from the 2017 excavation shows bones from small Atlantic cod (*Gadus morhua*) and large unidentified codfish (*Gadiformes*). At least one bone from a very large flatfish (*Pleuronectiformes*) is present. The fish bones are mainly ribs and fin bones, vertebrae, upper and lower jaws, and some other head bones. Additionally, several unspecified shell fragments were found in the material.

Overall, the species reflect a typical coastal Arctic environment and certainly indicate the exploitation of marine resources. Moreover, particular species of birds and sea mammals indicate that the dwelling could have been inhabited most of the year.

6 Dating

Besides the date for the seal bones, 2940–2570 calBC (4209±80, SPb-1561), we also obtained two conventional radiocarbon dates from the charcoal samples[1]. These reliable dates indicate the middle of the 3rd millennium calBC: 2570–2330 calBC (3950±45, SPb-2410) and 2620–2330 calBC (3972±50, SPb-2409)[2]. The first sample was procured from the floor area and the second one from the eastern part of the hearth.

The house site assemblage contains some diagnostic artefacts in terms of dating. First, it contains pottery sherds of the Palayguba type; this type was radiocarbon dated from 2500 to 1750 calBC (Zhul'nikov in press). Second, it contains polished slate arrowhead of the Sunderøy type. The earliest of these points dates back to approximately 2500–2100 calBC (Hesjedal et al. 2009: 418). Additionally, the type of antler harpoon head (hG2) points to the second half of the 3rd millennium calBC (Murashkin & Kiseleva 2018). In general, the types of artefacts indicate a period not earlier than 2500 calBC.

The house is situated at an elevation of 14,3–15,3 m in the Baltic elevation system. The relative sea level curve closest to the site was plotted for the Yarnyshnaya and Dalnezelenetskaya bays (Snyder et al. 1997). The relative sea level curve shows that the 15-metre level was exposed at around 5000 calBC, while the sea level in 2500 calBC was approximately 9–10 m above present levels.

7 Discussion

Due to the structural features revealed in the excavation, the dwelling at Kharlovka 1-6 is most similar to the so-called Gressbakken-type houses. Such houses are spread out along the Barents Sea coast and usually located in bays or fjords, though a few houses have been found on inland rivers in Norway (Simonsen 1963: 135–137, 162–168). Simonsen first described these dwellings as a specific type after extensive excavation work in Varangerfjord in the 1950s. Since then, such dwellings have been of much archaeological interest in Norway and later in Russia. However, this type of house has still not been clearly defined with a clear set of features. Common features of note include a large rectangular semi-subterranean chamber, a stone-framed double hearth placed along the longitudinal axis and multiple entrance passages. Frequently, the houses have wall-mounds containing refuse and raised platforms or annexes at both ends (Hood 2017; Schanche 1994: 3). The remains of chimney pipes have been recorded for several houses (Kolpakov et al. 2016: 180).

The vague definition of the "Gressbakken type" arises, among other things, from the small scale of excavations for most of the dwellings. In most cases, the excavation was limited to the floor area of the central chamber (or a part of it). Raised platforms and entrances/annexes were usually left outside the excavation areas and wall mounds were investigated via narrow trenches (for example Simonsen 1961: Fig. 114, 122; Schanche 1989: Fig. 3). Only a few houses at Gressbakken Nedre Vest (houses 3–4), Advik (houses b, j), Leirpollen, Bergeby, Bugøyfjord, Kalkillebukta and Høybukt can be regarded as more or less completely excavated (Fig. 1).

The most important features of Gressbakken houses (hearth structure, large size and depth of the chamber, platform at the end of the chamber) are represented in the Kharlovka 1-6 dwelling. In addition, the artefact assemblage is similar to the set of artefact types from the Gressbakken culture (phase). Thus, the dwelling at Kharlovka 1-6 is the seventh "classic Gressbakken" type house studied in Russia. Others include dwellings at the settlements of Grottug, Dvorovaya, Ust-Drozdovka 3 and Zavalishina 5 (Kolpakov et al. 2020; Carpelan & Ovsianikov 2013; Seitsonen 2006) (Fig. 1).

The majority of the Gressbakken-related ^{14}C dates in northern Norway fall between 2400/2200 and 1600 calBC, meaning they fall at the end of the Late Stone Age and Early Metal Period in the Norwegian chronological framework (Hood & Helama 2010; Olsen 1994; Schanche 1994). Radiocarbon dates for the Gressbakken-type houses on the Kola Peninsula were obtained from Ust-Drozdovka 3, Zavalishina 5 and Kharlovka 1-6. The large set of ^{14}C dates from house 5 at Ust-Drozdovka 3 fall between 2300 and 1500 calBC (Helskog et al. in press). Four ^{14}C dates were obtained from house 6 at Zavalishina 5 (Kolpakov et al. 2020: Table 2). Two of them are from the layers that filled the house chamber: 3490–2500 calBC (4327±150 BP, SPb-297) and 2470–1930 calBC (3754±100 BP, SPb-298). They were obtained from charred organic residues (food crusts) on ceramic sherds. However, they contradict each other. The first sample originates from horizon 6, and the second one from horizon 15. The second lies 60 cm lower than the first, but its dating is earlier than the first one. Therefore, the radiocarbon dates from Zavalishina 5 are not sufficiently reliable. At the same time, they date back to near the beginning of the Gressbakken tradition.

The Kharlovka 1-6 dates indicate a period from 2600 to 2300 calBC. They are earlier than the current dating for Gressbakken houses in northern Norway. It can be assumed that the Gressbakken-type dwellings in the eastern part of their distribution area, on the Murmansk coast, appeared several centuries earlier than in northeastern Norway. It should be noted that large semi-subterranean dwellings from the Late Neolithic period, with two (sometimes more) fireplaces situated along the long axis and entrances and/or annexes at the long ends of houses, have been found in Karelia (Zhul'nikov 2003: 59–63, 81; 2005: 85–96), Finland (Mökkönen 2011: 24–29; Pesonen 2002) and the Komi Republic (Karmanov 2017; Stokolos 1986, 1988). The similarities between the Late Neolithic ceramics of the Kola Peninsula and Karelia and the emergence of exotic materials (copper, amber) seem to reflect the direction of cultural influences from south to north. Perhaps some features of Gressbakken dwellings may reflect the influence of house-building traditions that existed in the vast taiga zone of Eastern Europe.

The faunal remains assemblage from the Kharlovka 1-6 house is quite similar to that from the Gressbakken-type houses in northern Norway (Hodgetts 1999; Olsen 1967; Renouf 1989; Schanche 1994). Osteological analysis of other faunal assemblages from the Kola Peninsula was conducted for Ust-Drozdovka 3 (Helskog et al. in press) and Zavalishina 5 (Kolpakov et al. 2016: 180–182). However, the data from Zavalishina 5 does not contain any identification of fish and bird bones. All three assemblages are dominated by marine mammal bones: harp seal, ringed seal, white whale, white-beaked dolphin and walrus. A small number of terrestrial mammal bones (wild reindeer, red fox/arctic fox, brown bear, Eurasian elk and Eurasian beaver) are represented. Just as with the Gressbakken-type houses in northern Norway, evidence exists of possible year-round residence (or for most of the year) at Ust-Drozdovka 3 and Kharlovka 1-6. The large number of harp seal bones (with a predominance of pups and juvenile animals) indicates active sea hunting in late winter and spring. Evidence of summer and autumn habitats is less abundant. The presence of migratory bird bones supports this conclusion.

The similarity between faunal materials and hunting equipments in most Gressbakken-type houses indicates that a common culture of sea hunter-gatherers has spread along the coast of the Norwegian Sea and Barents Sea, from Porsanger Fjord (Schanche 1994: 68–70) to St. Nose Cape (Kiseleva & Murashkin 2019; Kolpakov et al. 2016: 182–183).

8 Conclusion

In 2017–2019, the Kola Archaeological Expedition of the Institute for the History of Material Culture of RAS studied the remains of a house at Kharlovka 1-6 on the northern coast of the Kola Peninsula. The revealed construction has a number of features that are typical of so-called Gressbakken-type dwellings. According to the results of the excavations done on the Murmansk coast and in northern Norway, the area of distribution of such houses (and, in general, the Gressbakken culture) is approximately from Porsanger Fjord to St. Nose Cape. The radiocarbon dating, diagnostic types of stone and bone tools, and ceramics show that the house at Kharlovka 1-6 is one of the earliest houses of this type, dating to 2600–2300 calBC. Therefore, it may be that the tradition of Gressbakken-type houses emerged around the middle of the 3rd millennium calBC on the Kola Peninsula.

The faunal remains, which are dominated by marine mammals (especially harp seals) and seabirds, clearly indicate marine adaptation of the population living at the Kharlovka 1-6 site. This is also evidenced by fishing and hunting tools (harpoon and leister heads). The osteological analysis indicates that the site was exploited for most of the year, which is a consequence of the effective adaptation of the ancient population to life in the high Arctic.

Acknowledgements

The authors would like to thank all those who participated in the Kola Archaeological Expedition and who took part in the excavations. We also give our thanks to A. Luntsevich and N. Shashkov for assisting our fieldwork for many years. Special thanks go to S. Khavrin, E. Petrova and N. Volkova for analysis and consultations, to N. Vasil`yeva and V. Berezovskaya for restoration and to A. Malutina for the drawings and photos.

This work was carried out within the framework of the Programme of Fundamental Scientific Research of the State Academies of Sciences, State Assignment № 0184-2019-0002 and № 0160-2019-0044".

References

Archival sources

Gurina 1974 = Гурина, Н. Н. 1974. Отчет о работе Кольской археологической экспедиции за 1974 г. Архив ИИМК РАН, Ф. 35, 1974 г., д. 30

Kolpakov 2014 = Колпаков, Е. М. 2014. Отчет Кольской археологической экспедиции ИИМК РАН о разведках в Мурманской области 2014 г. Архив ИА РАН, Р-1, ед. хр. 45519

Shumkin 1978 = Шумкин, В. Я. 1978. Отчет о работе Разведочного отряда Кольской археологической экспедиции ЛОИА АН СССР за 1978 год. Архив ИИМК РАН, Ф. 35, 1978 г. д. 56

Unpublished sources

Helskog, K., Hood, B. C. & Shumkin, V. Ya. (in press). Dwelling Forms and Settlement Patterns on Russia's Kola Peninsula Coast, 2200-1500 cal. BC. Unpublished manuscript

Hodgetts, L. M. 1999. *Animal Bones and Human Society in the Late Younger Stone Age of Arctic Norway*. PhD theses, Durham University. Available at Durham E-Theses Online: http://etheses.dur.ac.uk/4491/

Hood, B. C. 2017. The Gressbakken Phase: «Complex» Hunter-Gatherers in Northern Norway? A Case-Study in Archaeological Interpretation. Unpublished manuscript.

Schanche, K. 1994. *Gressbakkentuftene i Varanger.*

Boliger og sosial struktur rundt 2000 f Kr. Doctoral thesis. University of Tromsø.

Zhul'nikov (in press) = Жульников А. М. О контактах населения с чирковской и асбестовой палайгубской керамикой. *Поволжская археология*.

Literature

Bronk Ramsey, C. 2009. Bayesian analysis of radiocarbon dates. *Radiocarbon*, 51(1): 337–360.

Gjessing, G. 1942. *Yngre steinalder i Nord-Norge*. Institutt for sammenlignende kulturforskning B XX-XIX, Oslo.

Carpelan & Ovsiannikov 2013 = Карпелан, К. & Овсянников, О. В. 2013. Саамское поселение первой четверти XVIII столетия на побережье Баренцева моря (Кольский полуостров, бухта Дворовая). In Брызгалов В. В. (ed.) *Лодия. 2010, № 7*: 261–326. Архангельск: Лодия, 2013.

Gurina 1997 = Гурина, Н. Н. 1997. *История культуры древнего населения Кольского полуострова*. Archaeologica Petropolitana 3.

Halén, O. 1994. *Sedentariness During the Stone Age of Northern Sweden in the Light of the Alträsket Site, c. 5000 B.C., and the Comb Ware Site Lillberget, c. 3900 B.C. Source Critical Problems of Representativity in Archaeology*. Acta Archaeologica Lundensia, Series in 4:o, 20. Stockholm.

Helskog, E. T. 1983. *The Iversfjord Locality. A Study of Behavioral Patterning during the Late Stone Age of Finnmark, North Norway*. Tromsø Museums Skrifter XIX. Tromsø: Tromsø Museum, Universitetet i Tromsø.

Hesjedal A., Damm C., Olsen B. & Storli I 1996. *Arkeologi på Slettnes. Dokumentasjon av 11.000 års bosetning*. Tromsø Museums Skrifter XXVI. Tromsø: Tromsø Museum.

Hesjedal, A., Ramstad, M. & Niemi, A. R. 2009. *Undersøkelsene på Melkøya. Melkøyaprosjektet – kulturhistoriske registreringer og utgravninger 2001 og 2002*. Tromura, Fellesserie 36. Tromsø: Tromsø Museum Universitetetsmuseet.

Hodgetts, L. & Rahemtulla, F. 2001. Land and sea: use of terrestrial mammal bones in coastal hunter-gatherer communities. *Antiquity* 75: 56–62.

Hood, B. C. & Helama, S. 2010. Karlebotnbakken Reloaded: Shifting the Chronological Significance of an Iconic Late Stone Age Site in Varangerfjord, North Norway. *Fennoscandia archaeologica XXVII*: 35–43.

Karmanov 2017 = Карманов В. Н. 2017. Новые данные по домостроительству населения крайнего северо-востока Европы в эпоху раннего металла. *Вестник Пермского университета*, Вып. 1(36): 26–40.

Kiseleva & Murashkin 2019 = Киселёва, А. М. & Мурашкин, А. И. 2019. Морская охота и рыболовство на побережье Северной Фенноскандии до рубежа эр. *Самарский научный вестник, Т*. 8, Вып. 2 (27): 171–179.

Kolpakov, E. M., Shumkin, V. Ya. & Murashkin, A. I. 2016. Early Metal Age Dwellings in Eastern Lapland: Investigations of the Kola Archaeological Expedition (IHMC) in 2004–2014. In P. Uino & K. Nordqvist (eds.) *New Sites, New Methods. Proceedings of the Finnish-Russian archaeological symposium Helsinki, 19–21 November, 2014*: 167–176. Iskos 21. Helsinki: The Antiquarian Society of Finland.

Kolpakov et al. 2020 = Колпаков, Е. М., Мурашкин, А. И. & Шумкин, В. Я. 2020. Древние жилища Восточной Лапландии – парадокс заполярной археологии. In В. В. Питулько & Н. В. Федорова (eds.) *Археология Арктики* 7: 278–299. Омск: Золотой тираж.

Krainov 1987 = Крайнов, Д. А. 1987. Фатьяновская культура. In О. Н. Бадер, Д. А. Крайнов & М. Ф. Косарев (eds.) *Эпоха бронзы лесной полосы СССР*: 58–75. Археология СССР. Москва: Наука.

Krenke 2019 = Кренке, Н. А. 2019. Радиоуглеродная хронология фатьяновской культуры. *Российская археология* 2: 110–116.

Mökkönen, T. 2011. *Studies on Stone Age Housepits in Fennoscandia (4000-2000 CAL BC). Changes in Ground Plan, Site Location, and Degree of Sedentism*. PhD dissertation, Helsinki: Unigraphia.

Murashkin & Kiseleva 2018 = Мурашкин, А. И. & Киселёва, А. М. 2018. Динамика развития костяного инвентаря Северной Фенноскандии (неолит – эпоха раннего металла) In Н. В. Федорова (ed.) *Археология Арктики* Вып. 5: 107–119. Омск: Золотой тираж.

Murashkin et al. 2019 = Мурашкин, А. И., Малютина, А. А. & Киселёва, А. М. 2019. Костяной и роговой инвентарь Кольского полуострова: типология, технология, трасология. *Записки Института Истории Материальной Культуры РАН*, 20: 85–103.

Niemi, A. & Oppvang, J. 2018. *Nyelv Nedre Vest. Utgravning av hustufter og boplass fra yngre steinalder*. Tromura, Kulturhistorie 46. Tromsø: Tromsø Museum – Universitetsmuseet, UiT Norges Arktiske Universitet.

Nordqvist, K. 2018. *The Stone Age of north-eastern Europe 5500–1800 calBC. Bridging the gap between the East and the West*. Acta Universitatis Ouluensis B Humaniora 160.

Nordqvist, K., Ikäheimo, J., Herva, V.-P. & Lahelma, A. 2011. Lillberget Reloaded: Re-Examining an Anomalous Comb Ware Site in Överkalix, Northern Sweden. *Fennoscandia archaeologica 28*: 71–78.

Olsen, B. 1994. *Bosetning og Samfunn i Finnmarks forhistorie*. Oslo: Universitetsforlaget.

Olsen, H. 1967. *Varanger-funnene IV. Osteologisk material. Innledning – fisk – fugl*. Tromsø Museums Skrifter VII: 4. Tromsø: Universitetsforlaget.

Pesonen, P. 2002. Semisubterranean Houses in Finland – a Review. In H. Ranta (ed.) *Huts and Houses. Stone Age and Early Metal Age Buildings in Finland*: 9–41. Helsinki: National Board of Antiquities.

Ramstad, M. 2006. Perler og mennesker 4000 f. Kr. Om miljøet rundt ravfunnene fra Finnmarks steinalder. In R. Barndon, S. M. Innselset, K. K. Kristoffersen, T. K. Lødøen (eds.) *Samfunn, symboler og identitet – Festskrift til Gro Mandt på 70-årsdagen*: 129–146. UBAS Nordisk 3.

Rankama, T. 1986. A Group of Late Stone Age Slate Arrowheads from Northernmost Finland and Norway.

In T. Edgren (ed.) *Studia præhistorica fennica C F Meinander septuagenario dedicata*: 127–139. Iskos 6. Helsinki: Suomen muinaismuistoyhdistys.

Reimer, P., Austin, W., Bard, E., Bayliss, A., Blackwell, P., Bronk Ramsey, C., Butzin, M., Cheng, H., Edwards, R., Friedrich, M., Grootes, P., Guilderson, T., Hajdas, I., Heaton, T., Hogg, A., Hughen, K., Kromer, B., Manning, S., Muscheler, R., Palmer, J., Pearson, C., van der Plicht, J., Reimer, R., Richards, D., Scott, E., Southon, J., Turney, C., Wacker, L., Adolphi, F., Büntgen, U., Capano, M., Fahrni, S., Fogtmann-Schulz, A., Friedrich, R., Köhler, P., Kudsk, S., Miyake, F., Olsen, J., Reinig, F., Sakamoto, M., Sookdeo, A., & Talamo, S. 2020. The IntCal20 Northern Hemisphere radiocarbon age calibration curve (0–55 cal kBP). *Radiocarbon* 62: 725–757.

Renouf, M. A. P. 1989. *Prehistoric Hunter-Fishers of Varangerfjord, Northeastern Norway. Reconstruction of Settlement and Subsistence During the Younger Stone Age*. BAR International Series 487.

Schanche, K. 1989. Nye funn fra yngre steinalder i Varanger. *Viking LII*: 53–71.

Seitsonen, O. 2006. Petsamo Maattivuono Rotojoki: two Late Stone Age dwellings excavated by Sakari Pälsi in 1929. In V.-P. Herva (ed.) *People, Material Culture and Environment in the North: Proceedings of the 22nd Nordic Archaeological Conference, University of Oulu, 18–23 August 2004*: 226–237. Studia Humaniora Ouluensia 1.

Simonsen, P. 1961. *Varangerfunnene II. Fund og udgravninger i Pasvikdalen og på den østlige fjordstrand.* Tromsø Museum Skrifter VII: 2. Tromsø: Tromsø museum.

Simonsen, P. 1963. *Varangerfunnene III. Fund og utgravninger i Pasvikdalen og ved den østlige fjørdstrand.* Tromsø Museums Skrifter VII: 3. Tromsø: Universitetsforlaget.

Snyder, J. A., Forman, S. L., Mode, W. N. & Tarasov, G. A. 1997. Postglacial relative sea-level history: sediment and diatom records of emerged coastal lakes, north-central Kola Peninsula, Russia. *Boreas* 2: 329–346.

Stokolos 1986 = Стоколос, В. С. 1986. *Древние поселения Мезенской долины*. Москва: Наука.

Stokolos 1988 = Стоколос, В. С. 1988. *Культуры эпохи раннего металла Северного Приуралья*. Москва: Наука.

Zhul'nikov 1999 = Жульников, А. М. 1999. *Энеолит Карелии (памятники с пористой и асбестовой керамикой)*. Петрозаводск: Карельский научный центр РАН.

Zhul'nikov 2003 = Жульников, А. М. 2003. *Древние жилища Карелии*. Петрозаводск: Скандинавия.

Zhul'nikov 2005 = Жульников, А. М. 2005. *Поселения эпохи раннего металла Юго-Западного Прибеломорья*. Петрозаводск: Паритет.

Zhul'nikov, A. M., 2008. Exchange of amber in northern Europe in the III Millennium BC as a factor of social interactions. *Estonian Journal of Archaeology* 12 (1): 3–15.

Notes

1 Wood species was not identified.
2 All dates were calibrated by OxCal v4.4.2 (Bronk Ramsay 2009); r5; IntCal 2020 curve; Atmospheric data from Reimer et al. 2020.

The Use of Mineral Raw Materials in Lithic Knapping in Stone Age Northern Ostrobothnia, Finland

Teemu Mökkönen & Kerkko Nordqvist

Teemu Mökkönen, Eastern and Northern Finland / Cultural Environment Services, The Finnish Heritage Agency. P.O. Box 913, 00101 Helsinki, Finland, e-mail: teemu.mokkonen@gmail.com
Kerkko Nordqvist, Department of Cultures, University of Helsinki, Unioninkatu 38, 00014 Helsingin yliopisto, Finland; Department of Archaeology, University of Tartu. Jakobi 2, 51014, Tartu, Estonia, e-mail: kerkko.nordqvist@helsinki.fi

Abstract

Although quartz predominates in Stone Age knapped assemblages in Finland, the selection of raw materials was not similar throughout the period. This article takes a closer look at the choice of lithic raw materials in Northern Ostrobothnia (Finland) during the 6th–3rd millennia calBC. It demonstrates that changes in lithic raw materials and other cultural development, including pottery production, do not always go hand in hand. A comparison of Northern Ostrobothnia with the Lake Saimaa area shows that there are certain similarities in lithic assemblages (raw material use and technology) during some cultural phases despite the different local raw material base, although diverging trends and regional variation can also be identified.

1 Introduction

The enormous quantity of (vein) quartz that makes up the majority of lithic assemblages is probably the main reason why the selection of Stone Age raw materials is a marginally studied topic in Finland. Only two short periods with a large amount of imported flint – the pioneer settlement right after the last glaciation in the 9th millennium calBC and the Typical Comb Ware period in 3900–3400 calBC – have aroused more interest (Hertell & Tallavaara 2011; Jussila et al. 2012; Kankaanpää & Rankama 2014; Kinnunen et al. 1985; Mökkönen et al. 2017a; Rankama et al. 2006; Vuorinen 1982). Quartz is often considered a profoundly bulk material. Its irregular internal structure makes it a difficult material to control when knapping (e.g. Rankama et al. 2006: 249–250), and quartz assemblages with a lot of debris and few diagnostic artefacts are common.

Due to the lack of studies summarising the regional use of various raw materials, the overall picture of Stone Age lithics in Finland is perceived as being rather monotonous. Recent papers by the present authors have shown variation in the selection of raw materials and detected both temporal and geographical trends (Mökkönen & Nordqvist 2016; Mökkönen et al. 2017a). This article is a direct continuation of these studies. It provides an overview of the development of lithic raw materials used in knapping in the province of Northern Ostrobothnia (Finland) from the 6th to the 3rd millennium calBC (Fig. 1). In the following, we will focus on the general characteristics of the lithic raw material use, as well as their temporal and technological changes, and place them

Figure 1. The location of the studied sites. The solid line shows the borders of the province of Northern Ostrobothnia, Finland. In the Stone Age, all sites were located on the seacoast (except for no. 1). Sites: 1 – Tervaniemi, Ahola; 2 – Vepsänkangas; 3 – Veskankangas; 4 – Tainiaro; 5 – Latokangas; 6 – Kivimaa; 7 – Kierikin sorakuoppa; 8 – Kierikinkangas; 9 – Kauniinmetsänniitty 1; 10 – Kuuselankangas; 11 – Purkajasuo-Korvala; and 12 – Hangaskangas E. Rivers: Ke – Kemijoki; Ki – Kiiminkijoki; and Ou – Oulujoki. Map: T. Mökkönen, 2020, vector base map by Natural Earth.

in a wider context by comparing the results with the situation in the Lake Saimaa area (eastern Finland).

The analysed assemblages are divided into five phases that roughly correspond to the general Finnish archaeological periodisation and pottery sequence (see Nordqvist & Mökkönen 2017) (Table 1, Appendix 1). *Phase 1: 6000–5000 calBC* encompasses the Late Mesolithic and the first centuries of the Neolithic (defined in the research area by the appearance of pottery from 5300 calBC), *Phase 2: 5000–4000 calBC* corresponds to the time of the earliest pottery traditions (Säräisniemi 1 Ware, Sperrings 1 and 2 Wares), *Phase 3: 4000–3500 calBC* covers the Typical Comb Ware period, *Phase 4: 3500–3000 calBC* includes the asbestos- and organic-tempered Kierikki and Pöljä Wares as well as local late variants of the Comb Ware tradition, and *Phase 5: 3000–2000 calBC* matches with the Late and Final Neolithic, but is not associated with any pottery in the studied assemblages. This division was also used in our previous studies.

The province of Northern Ostrobothnia is located on the eastern coast of the Bothnian Bay in the Baltic Sea but extends to the Finnish-Russian border. Its geography is characterised by a relatively short coastal strip, which is connected to the vast inland areas by an extensive network of waterways. The most common soil is moraine, but there are several different bedrock areas in the region. In addition, the geologically diverse area of southern and central Lapland (Silvennoinen 1998) is situated to the north of it. The latter environment set the stage for the development of a particular regional lithic industry, whose influence was also felt in the research area. A short excursion into this northern 'slate culture' can serve as a prologue to our study.

	calBC	Sites	Lithics (g/pcs)	Lithics per m^2 (g/pcs)
Phase 1	6000–5000	1	5911/3198	19/10
Phase 1+2	6000–4000	2	69 988/13 559	145/28
Phase 2	5000–4000	1	133 562/14 586	266/36
Phase 2+3	5000–3500	1	18 144/4 065	25/6
Phase 3	4000–3500	4	61 530/12 552	71/14
Phase 3+4	4000–3000	1	50 023/21 849	95/41
Phase 4	3500–3000	1	22 209/12 467	126/71
Phase 5	3000–2000	1	2 078/1 100	43/23
Total		12	363 445/83 376	

Table 1. Chronological phases used in the study and the volume of lithics per phase (for data, see Appendices 1 and 2).

2 Close to the slate zone

In southern Lapland, slates, in addition to quartz, were a characteristic part of the Stone Age raw material kit for a long time. The term 'slate' used by archaeologists (and also retained here) is imprecise and can cover the various (meta)tuffs, phyllites, shales, schists, quartzites, etc. available in outcrops to the west of the Finnish-Swedish border (see Silvennoinen 1998; Sarala 2012). One key area with significant local raw material deposits is known to be located in the municipality of Tervola in the lower Kemijoki valley, and it is believed to have been a production centre for the so-called Bothnian ground stone tools (Kotivuori 1996: 67; see also Eranti 2020; Huurre 1983: 105; Kelloniemi 2020).

The use of local slate outcrops gained momentum in the 7th millennium calBC at the latest and intensified further between the 6th and 4th millennia calBC when new polished artefact types appeared (Huurre 1983: 100–109, 160–168; Kotivuori 1996: 57–63; Nunez 1998; see Hallgren 2008: 255–260 for the corresponding 'slate culture' in northern Sweden). It has been estimated that during the most intensive phase of slate utilisation in southern Lapland the knapping of quartz and slate was equally common, and quartzite was used, too (Kotivuori 1996: 63, 67). Still, knapped slate artefacts are poorly studied, and formal (retouched) artefacts are rare (see below): slate raw materials often fracture neatly along smooth cleavage plains, resulting in sharp flakes that can be used as tools without further processing.

Pottery technology, which began to spread after 5300 calBC, was initially rather weakly rooted in the research area and rarely crossed the present Finnish-Swedish border (but see Norberg 2008: 23–24 for more western encounters). Typically, only a small volume of pottery is found at the sites (Huurre 1983: 137, 146; Nordqvist 2018: 94–97), at least when compared to further south in Finland, and many areas occupied by the 'slate culture' remain aceramic. Thus, our research area – Northern Ostrobothnia – can be described as a transition zone located at the crossroads of several routes and cultural traditions; this also characterizes it in the later phases of prehistory.

3 Analysed sites and materials

The studied materials come from 12 settlement sites excavated between 1984 and 2012 (Fig. 1, Appendix 1), and include 83 376 lithic objects with a total weight of over 363 kg (Table 1). These materials have not been used in previous quantitative lithic studies and were analysed by the authors at the premises of the Finnish Heritage Agency.

Recently excavated single-period contexts (assemblages) with a sufficient level of documentation and find recovery (sieving) were preferred in order to facilitate reliable comparisons between phases, sites and regions. The analysis of the materials focused on recording a predefined set of variables and was performed with the naked eye (for the analytical procedures, see Mökkönen & Nordqvist 2016: 44–45).

All the studied localities are situated in Northern Ostrobothnia, with the exception of the Tainiaro site, which is located a few kilometres away in Lapland province. During the Stone Age, these were all coastal sites except for the Tervaniemi, Ahola site, which was located inland. Due to the rapid land uplift in the area (current rate of ca. 7 mm/year), the settlement sites following the changes of the seashore were typically occupied for only a limited span of time. This creates good conditions for the formation of short-lived settlement sites and for the recognition of temporal changes in material culture. Based on radiocarbon data, the analysed sites have been inhabited for about two or three centuries, but less than 500 years (Appendix 2). The dates of the Latokangas and Kauniinmetsänniitty 1 sites indicate a longer settlement history, even if the typologically diagnostic materials belong only to the older settlement phase (see also Pesonen 2013a). Wood charcoal dates with large standard deviations place the Tervaniemi, Ahola site only roughly to the 6th millennium calBC.[1]

The material covers the period from the 6th to the 3rd millennium calBC, and the studied assemblages also represent all the main pottery types encountered in the area (Appendix 1). Two of the early sites, Tervaniemi, Ahola and Veskankangas, do not contain pottery, although the latter is already coeval with the early use of ceramics in the area. The youngest assemblage – the Hangaskangas E site, which dates to the Final Neolithic – also lacks pottery.

Due to the long period of time studied here, the phases are mostly represented by only one or two clean (unmixed) assemblages, and broad dates link some sites to two phases. Therefore, the results and temporal trends presented in this article for Phases 2 through 4 can be considered preliminary, while the oldest and youngest phases should be approached with caution.

4 Results and discussion

4.1 Raw material selection

Quartz, which is one of the most abundant locally available minerals in Finland, was the key material in Northern Ostrobothnia throughout the Stone Age (Fig. 2, Appendix 3). Its share of all lithics is typically around 80% (here and below, per cent by weight), and at some sites it can reach nearly 100%. Quartz is followed by slate, the proportion of which can reach 30–90% of all lithics, especially in the early phases. Quartzite is present in some amount throughout the Stone Age, while other raw materials (flint and other stones[2]) are found in smaller quantities.

Phase 1 is only represented by the quartz-dominated (>95%) inland site of Tervaniemi, Ahola. The Phase 2 sites illustrate the lithic use at a time when the earliest pottery was spreading in the area. The two northernmost sites of the study, Veskankangas and Tainiaro, are also the sites most strongly characterised by slate. In contrast, the other two Phase 2 sites (Latokangas and Vepsänkangas), located approximately one hundred kilometres south of the slate-rich sites, are much more consistent with the quartz-dominated profiles of the younger sites in the research area. This indicates that the geographical proximity to the slate outcrops of southern Lapland was an important factor in their utilisation and, together with the properties of the raw material itself, contributed to the way and volume of lithic working (see e.g. Manninen & Knutsson 2014). Slate artefacts from the analysed sites include several items typical for the industry of southern Lapland (and especially sites containing the earliest pottery; see Kotivuori

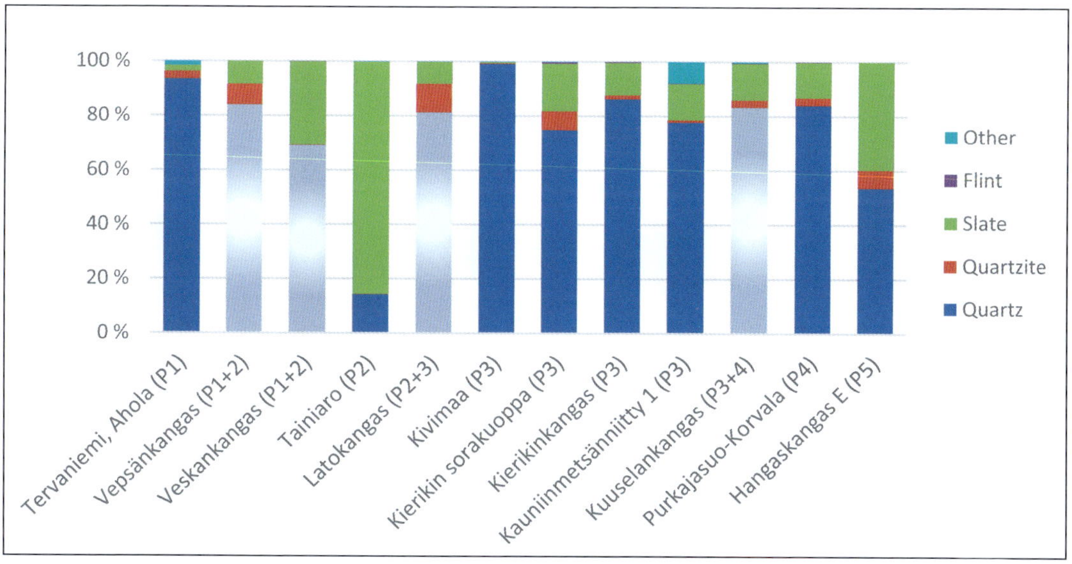

Figure 2. Percentage distribution of the lithic raw materials by weight at the analysed sites. Materials dating to two phases are indicated by a soft gradient colour on the quartz bars. The phases are shown in parentheses.

1996: 61–62, 93), such as slightly retouched or polished knives, tanged scrapers, points and blades, as well as retouched flakes and reused fragments of ground stone tools (Fig. 3). They show that the border of different lithic traditions, one relying on quartz and the other intensively using slates, runs in the 5th millennium calBC in the northern part of Northern Ostrobothnia.

The lithic profiles of the Phase 2 sites also suggest that no particular lithic technology spread alongside the early pottery traditions (Säräisniemi 1 and Sperrings 1 Wares), and that the appearance of ceramics did not change the local traditions of lithic raw material use. This is similar to the situation observed in the Lake Saimaa area (Mökkönen & Nordqvist 2016) and elsewhere in northeastern Europe (Gerasimov & Kriiska 2018: 307; Núñez 1990: 41). The Veskankangas site, which is rather rich in slate, also hints that some groups may not have been directly involved in the contacts that spread the early pottery, or for other reasons chose not to adopt or use ceramics in all contexts.

Our results also confirm previous observations that the appearance of Typical Comb Ware (Phase 3) did not lead to large-scale import of flint in Northern Ostrobothnia (see Vuorinen 1982: 81–82; also Mökkönen et al. 2017a) (Fig. 2). The largest share of flint is present at the Kierikin sorakuoppa site, but the amount – 1% – is insignificant compared to the Lake Saimaa area, where the corresponding figure averages around 30% (Mökkönen & Nordqvist 2016; Mökkönen et al. 2017a). In addition, flint often came to Northern Ostrobothnia in the form of ready-made artefacts and not as a raw material, as in the south. In the Lake Saimaa area, the intensified use of flint also triggered the use of local rock crystals (on average 17% of the analysed catalogue sub-numbers; Mökkönen et al. 2017a), but this is not clearly visible in Northern Ostrobothnia (Appendix 3). An increased content of high-quality quartzes is only recorded at the aforementioned Kierikin sorakuoppa site (8%) and the fully quartz-dominated Kivimaa site (10%); the latter is the southernmost of the analysed locations and has direct water connections to central and eastern Finland. Thus, even if the emergence of Typical Comb Ware was associated with a new lithic technology in

Figure 3. Slate artefacts from the Tainiaro (a–h, m) and Veskankangas (i–l) sites: tanged scraper (a); axes/adzes (b, e, f); knives (c, d, h); awls, points and blades (g, l, m); scraper (k); scrapers retouched from fragments of polished artefacts (j, l). Tools that take advantage of the naturally sharp edges of split pieces with minimal retouch are typical of the slate industry. a – KM 25797: 634; b – KM 25797: 2062; c – KM 24925: 933; d – KM 22398: 619; e – KM 22398: 375; f – KM 25797: 1701; g – KM 22398: 227; h – KM 22398: 546; i – KM 24928: 1261; j – KM 24928: 783; k – KM 27365: 250; l – KM 24423: 324; m – KM 22398: 67. Photos: T. Mökkönen, 2014, illustration K. Nordqvist, 2020.

the Lake Saimaa area, the development in Northern Ostrobothnia followed a slightly different path, which is also illustrated by other material culture (Mökkönen & Nordqvist 2018).

The transition to Phase 4 is marked by the appearance of asbestos- and organic-tempered potteries (Kierikki and Pöljä Wares). However, in Northern Ostrobothnia the lithic raw material profiles remain practically unchanged in comparison with the preceding phase (only the few high-quality quartzes disappear; Appendix 3). In the Lake Saimaa area, this period is characterised by a sharp decline in the flint import and a return to the almost complete predominance of (vein) quartz (Mökkönen & Nordqvist 2016; Mökkönen et al. 2017a). Again, despite general similarities within the ceramic-centric narrative of cultural development, the local backgrounds and development trajectories differ in these areas.

Material dating to the 3rd millennium calBC (Phase 5) may indicate an increase in

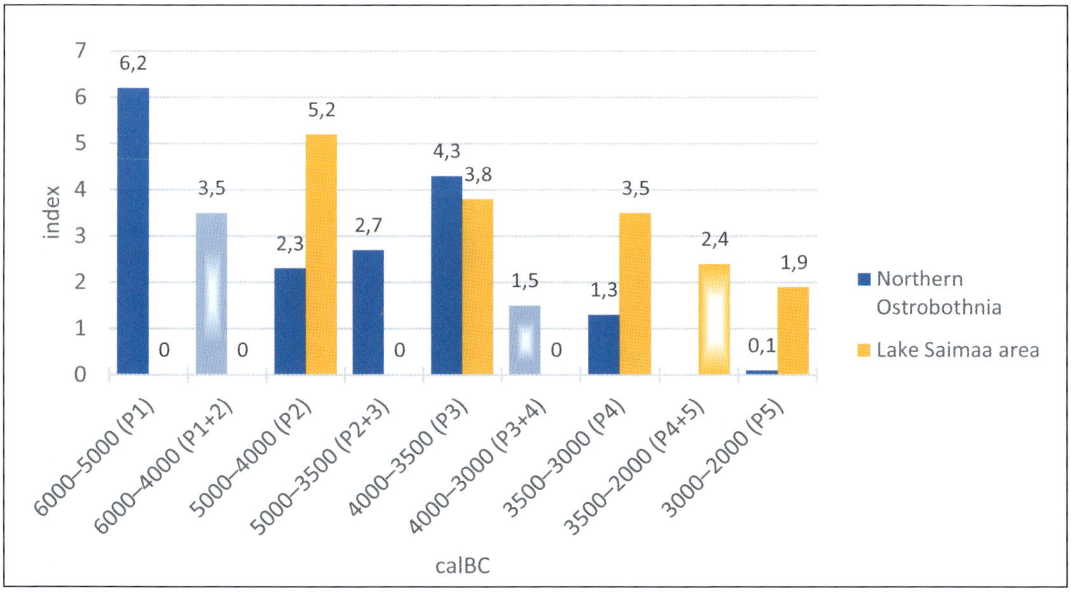

Figure 4. The quartz tool-to-flake ratios in Northern Ostrobothnia (site n 12) and the Lake Saimaa area (n 18) according to phases (pcs); index values are the average values for each period multiplied by 100, with the exception of Northern Ostrobothnian Phases 1, 2, 2+3, 3+4, 4, 5, as well as Phase 4 of the Lake Saimaa area, each of which is represented by only one site. The higher the value, the greater the proportion of tools in relation to debitage. Materials dating to two phases are indicated by a soft gradient colour on the quartz bars. The phases are shown in parentheses. Data for Lake Saimaa after Mökkönen & Nordqvist 2016.

slate use at the very end of the Stone Age in the River Oulujoki area (Fig. 2), but there is too little material for reliable conclusions.

4.2 Technological choices and quartz

The largest proportion of pieces with cortex is present at the early sites (Phases 1+2 and 2), on average 12% of the analysed catalogue sub-numbers (Appendix 4). The high frequency of cortex indicates firstly that quartz pebbles were used as raw material, and secondly that early phases (initial reduction) of the knapping sequence took place at the site. At younger sites, the share of cortex pieces remains 9% and lower, with the exception of the Kivimaa (18%, Phase 3) and Hangaskangas E (17%, Phase 5) sites. Thus, the trend in Northern Ostrobothnia is quite the opposite of the Lake Saimaa area, where the highest values (up to 12–14%) are reached in Phases 4 and 5, and the lowest (4–6%) in Phases 2 and 3 (Mökkönen & Nordqvist 2016).

The tool-to-flake ratio (Fig. 4) is generally decreasing over time. This indicates that the number of formal quartz tools in relation to debitage found at the sites is decreasing, i.e. that the knapping done at the sites (or in our analysed areas) has increased in volume and/or included more stages of the entire reduction process. Likewise, preferences for the production of formal tools and their discard patterns may have changed. However, the apparent rise in the tool-to-flake ratio in Phase 3 indicates that during the Typical Comb Ware period a larger part of the knapping sequence took place outside of the settlements. Together with a simultaneous decrease in the share of cortex pieces, this can also indicate a shift in the raw material base from pebbles to vein quartz or, alternatively, a pretreatment of pebbles (cores) elsewhere.

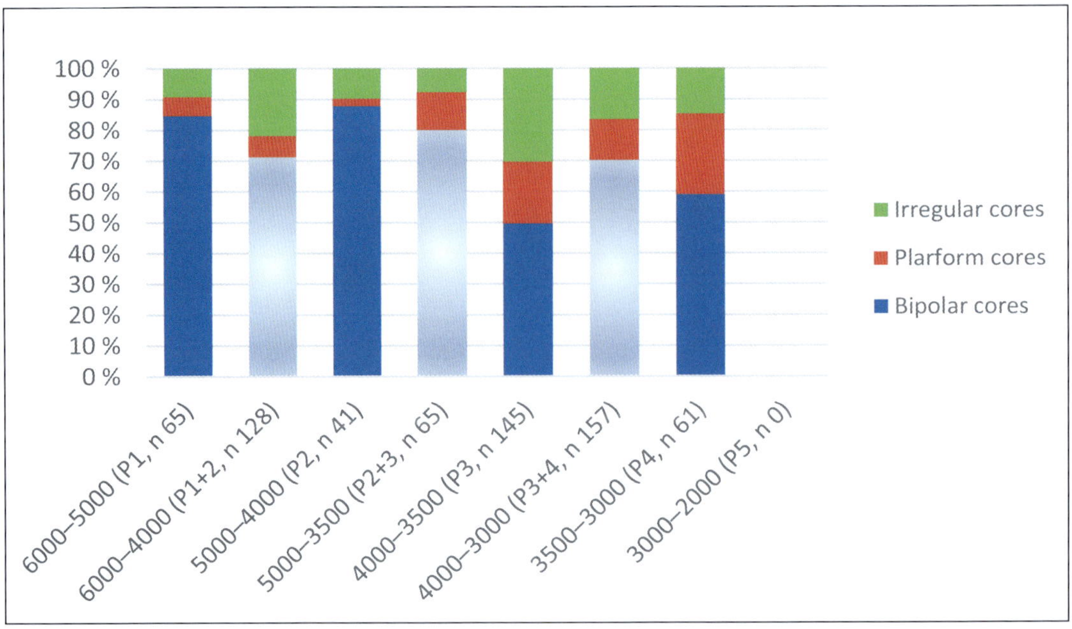

Figure 5. Percentage distribution of the quartz core types. 'Irregular cores' include undefined, fragmentary or exhausted cores. Materials dating to two phases are indicated by a soft gradient colour in the bipolar bar. The phases and number of cores (n) are shown in parentheses.

Based on the core data, bipolar technology prevailed in quartz-dominated knapping throughout the Stone Age (Fig. 5, Appendix 4), but its ratio to platform flaking varied in different phases[3]. In Northern Ostrobothnia, the largest share of bipolar cores is recorded in Phases 1–2, when they account for up to >90% of the quartz cores. Their proportion deceases during Phases 3 and 4 and usually remains between approximately 70 and 85%, although some assemblages with a much higher proportion of platform cores (especially Kierikin sorakuoppa, >40%) indicate site-level variations in technology during Phase 3. However, the general trends differ from those in the Lake Saimaa area (Fig. 6). There, the share of bipolar quartz cores in Phase 2 is moderate (ca. 55%) and increases in Phase 3 (about 80%, but just ca. 15% in flint cores) and after it, reaching up to 90% in Phases 4 and 5 (Mökkönen & Nordqvist 2016). These changes are fairly consistent with the changes in the use of quartz raw materials described above.

The differences in the knapped lithics in Northern Ostrobothnia and the Lake Saimaa area reflect not only local resource bases, but also regional traditions, development trajectories and contact networks. The slate-based industry prevailing in the north during Phases 1–2 (Fig. 3) is a case in point and attests to the importance of the local raw material base and the deep roots of local lithic working traditions. At the beginning of the 4th millennium calBC (Phase 3), the introduction of Typical Comb Ware in both areas was accompanied by changes in the lithic knapping performed at the sites (e.g. a high tool-to-flake ratio and a low proportion of cortex pieces). It is noteworthy that in Northern Ostrobothnia there was no broad introduction of flint-based platform flaking or large-scale use of high-quality quartzes. At the same time, some of the Phase 3 sites, primarily Kierikin sorakuoppa, but also Kivimaa, exhibit qualities that are more reminiscent of Lake Saimaa and evidence of differences in development

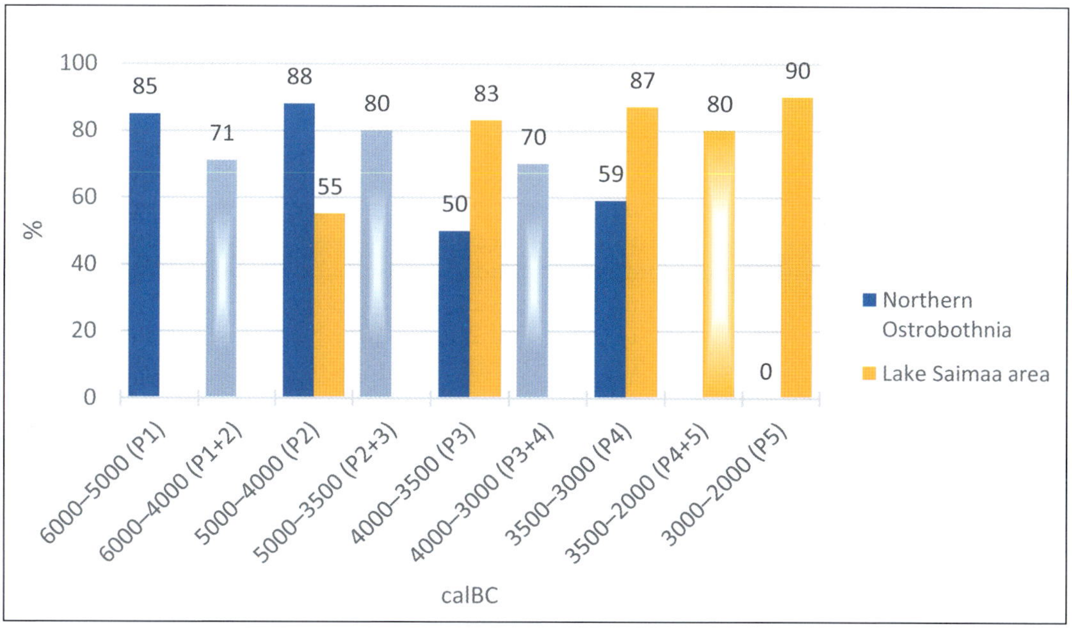

Figure 6. Proportion of bipolar cores from quartz corers in Northern Ostrobothnia and the Lake Saimaa area. Materials dating to two phases are indicated by a soft gradient colour. Data for Lake Saimaa after Mökkönen & Nordqvist 2016.

and contacts within the research area (see also Mökkönen & Nordqvist 2018). The transition to Phase 4 is signalled by asbestos- and organic-tempered potteries, as well as the disappearance of flint and high-quality quartzes and an overall quantitative increase in quartz lithics in both areas (pcs/g per excavated m^2; see Table 1 and Mökkönen & Nordqvist 2016). Despite this, the lithic profiles remain different (Figs. 4 and 6), and the changes in Northern Ostrobothnia are more subtle than in the Lake Saimaa area, where the later 4th millennium calBC is characterised by a sharp turn to the bipolar reduction of vein quartz.

5 Concluding remarks

This paper presents an overview of the lithic raw material use by hunter-fisher-gatherers who inhabited the coastal areas of Northern Ostrobothnia between the 6th and 3rd millennia calBC. The assemblages are mainly dominated by quartz and appear monotonous at first. Nevertheless, a closer look at the basic metrics (the proportion of pieces with cortex, the tool-to-flake ratio, the frequency of different core types) and a comparison with similar data from the Lake Saimaa area reveal different but also partially overlapping developments in these two areas.

During the early phases (1–2, 6th–5th millennia calBC), the slate industry was still prevailing in the northernmost part of the research area but did not reach the River Kiiminkijoki area (the Latokangas and Vepsänkangas sites); corresponding ground slate tools found further south are interpreted as being imported from the north (Äyräpää 1950: 35). This slate industry was basically aceramic, and even if Phases 1–2 coincide with the introduction of early pottery technology (Säräisniemi 1 and Sperrings 1 Wares), the research area is near the northwestern limit of its distribution. There is no indication that the arrival of pottery craft would have directly influenced the local lithic working traditions.

Except for the slate-rich northern sites, quartz prevails at all sites studied. In contrast to the Lake Saimaa area, the appearance of Typical Comb Ware (Phase 3) did not result in a sharp increase in imported flint and platform flaking. However, there are certain indicators in general knapping practices (e.g. the tool-to-flake ratio, the proportion of pieces with cortex) and especially at the site level (see the Kierikin sorakuoppa site with more flint, high-quality quartzes and platform flaking) that link the two areas. In Phase 4, similarities can be found in the general quantitative increase of quartz lithics, but here too the lithic profiles are different, as the Lake Saimaa area shows a much more acute turn towards bipolar working of (vein) quartz compared to the previous phase.

Current material from Northern Ostrobothnia shows that, despite the similar broad outlines of general cultural development, the differences between the analysed assemblages and periods are quite large. These can often be explained by the geographical location of the sites, both in relation to raw material sources and the direction of communication routes (waterways) and social ties. The slate industry is an example of a phenomenon based on links to the north, while the emergence of pottery, as well as changes during Phases 3 and 4, are associated with connections to the east and southeast. In Northern Ostrobothnia, however, there are no such abrupt changes that occurred at the beginning of Phase 3 in the Lake Saimaa area and were probably associated with the arrival of a new population (see Mökkönen et al. 2017b: 185; Nordqvist 2018: 101–103). Rather, changes in material culture, technology and lifestyle are the result of other forms of mobility and processes bringing new features into the research area that was at the intersection of different cultural areas and impulses.

Acknowledgements

The material of this study was collected in the project 'The Use of Materials and the Neolithisation of North-Eastern Europe (c. 6000–1000 BC)' funded by the Academy of Finland (2013–2017, decision 269066) and based at the University of Oulu, Finland. Part of the final writing was funded by the Helsinki University Humanities Project (KN).

References

Unpublished sources

Eranti, O. 2020. Geochemical comparison of two Tervola volcanite deposits as plausible stone tool material sources in Kemijoki. MA Thesis, University of Helsinki.

Kelloniemi, A. 2020. Sädekiviliuskeen olemus. Liuskekiven käyttö varhaisneoliittisella Simo-Tervola -alueella, tapaustutkimuksena Tainiaro. MA Thesis, University of Oulu.

Koivisto, S. 1997. Ylikiiminki Vepsänkangas. Kivikautisen asuinpaikan kaivaus 28.5.–27.6.1997. Research report, Finnish Heritage Agency, Helsinki.

Koivisto, S. 1998. Ylikiiminki Vepsänkangas. Kivikautisen asuinpaikan kaivaus 1998. Research report, Finnish Heritage Agency, Helsinki.

Pesonen, P. 2000. Yli-Ii [43] Kuuselankangas: Kivikautisen asuinpaikan kaivaus. Research report, Finnish Heritage Agency, Helsinki.

Pesonen, P. 2013b. Oulu Hangaskangas E. Kivi- ja pronssikautisen asuinpaikan kaivaus 14.6.–20.7.2012. Research report, Finnish Heritage Agency, Helsinki.

Raike, E. 1994. Taivalkoski 37 Tervaniemi. Mesoliittisen ajan asuinpaikan kaivaus 1994. Research report, Finnish Heritage Agency, Helsinki.

Literature

Äyräpää, A. 1950. Skifferspetsarna i östra Fenno-Skandien. *Finskt Museum* 57 (1950): 5–52.

Bronk Ramsey, C. 2009. Bayesian Analysis of Radiocarbon Dates. *Radiocarbon* 51(1): 337–360.

Gerasimov, D. & Kriiska, A. 2018. Early-Middle Holocene Archaeological Periodization and Environmental Changes in the Eastern Gulf of Finland: Interpretative Correlation. *Quaternary International* 465: 298–313.

Hallgren, F. 2008. *Identitet i praktik. Lokala, regionala och överregionala sociala sammanhang inom nordlig trattbägarkultur*. Coast to Coast-book 17.

Hertell, E. & Tallavaara, M. 2011. High Mobility of Gift Exchange – Early Mesolithic Exotic Chipped Lithics in Southern Finland. In T. Rankama (ed.) *Mesolithic Interfaces – Variability in Lithic Technologies in Eastern Fennoscandia*: 11–41. Monographs of the Archaeological Society of Finland 1.

Huurre, M. 1983. *Pohjois-Pohjanmaan ja Lapin esihistoria*. Pohjois-Pohjanmaan ja Lapin historia I. Kuusamo: Pohjois-Pohjanmaan maakuntaliiton ja Lapin maakuntaliiton yhteinen historiatoimikunta.

Jussila, T., Kriiska, A. & Rostedt, T. 2012. Saarenoja 2. An Early Mesolithic Site in South-Eastern Finland: Preliminary Results and Interpretations of Studies Conducted in 2000 and 2008–10. *Fennoscandia archaeologica* XXIX: 3–27.

Kankaanpää, J. & Rankama, T. 2014. Fast or Slow Pioneers? A View from Northern Lapland. In F. Riede & M. Tallavaara (eds.) *Lateglacial and Postglacial Pioneers in Northern Europe*: 147–159. BAR International Series 2599.

Kinnunen, K., Tynni, R., Hokkanen, K. & Taavitsainen, J.-P. 1985. *Flint Raw Materials of Prehistoric Finland: Rock types, Surface Textures and Microfossils*. Geological Survey of Finland, Bulletin 334.

Kotivuori, H. 1996. Pyytäjistä kaskenraivaajiksi. Rovaniemen asutus n. 6000 eKr.–1300 jKr. In M. Saarnisto, H. Kotivuori, J. Vahtola & M. Enbuske (eds.) *Rovaniemen historia vuoteen 1721*: 36–125. Rovaniemi: Rovaniemen kaupunki, Rovaniemen maalaiskunta & Rovaniemen seurakunta.

Manninen, M. A. & Knutsson, K. 2014. Lithic Raw Material Diversification as an Adaptive Strategy – Technology, Mobility, and Site Structure in Late Mesolithic Northernmost Europe. *Journal of Anthropological Archaeology* 33: 84–98.

Mökkönen, T. & Nordqvist, K. 2016. Quantifying Mineral Raw Materials in Neolithic Knapped Tool Production in the Lake Saimaa Area, Finnish Inland. In P. Uino & K. Nordqvist (eds.) *New Sites, New Methods. Proceedings of the Finnish-Russian Archaeological Symposium Helsinki, 19–21 November, 2014*: 41–58. Iskos 21.

Mökkönen, T. & Nordqvist, K. 2018. Kierikki Ware and the Contemporary Neolithic Asbestos- and Organic-tempered Potteries in North-East Europe. *Fennoscandia archaeologica* XXXIV: 83–116.

Mökkönen, T., Nordqvist, K. & Herva, V.-P. 2017a. Beneath the Surface of the World: High Quality Quartzes, Crystal Cavities, and Neolithization in Circumpolar Europe. *Arctic Anthropology* 54 (2): 94–110.

Mökkönen, T., Nordqvist, K. & Herva, V.-P. 2017b. Changes in Neolithic Lithic Raw Materials in Eastern Finland: Indications of Changing Contact Networks. In D.V. Gerasimov (ed.) *Cultural Processes on the Circum-Baltic Space in the Early and Middle Holocene. Papers of the International Scientific Conference Dedicated to the Memory of Vladimir I. Timofeev, Saint-Petersburg, Russia, April 26–28, 2017 / Культурные процессы в циркумбалтийском пространстве в раннем и среднем голоцене: Доклады международной научной конференции, посвященной 70-летию со дня рождения В.И. Тимофеева, Санкт-Петербург, Россия, 26–28 апреля 2017 г.*: 181–186. Saint-Petersburg: MAE RAS.

Norberg, E. 2008. *Boplatsvallen som bostad i Norrbottens kustland 5000 till 2000 före vår tideräkning. En studie av kontinuitet och förändringar*. Studia Arcaheologica Universitatis Umensis 23.

Nordqvist, K. 2018. *The Stone Age of North-Eastern Europe 5500–1800 calBC. Bridging the gap between the East and the West*. Acta Universitatis Ouluensis, B Humaniora 160.

Nordqvist, K. & Mökkönen, T. 2017. Periodization of the Neolithic and Radiocarbon Chronology of the Early Neolithic and the Beginning of Middle Neolithic in Finland. *Documenta Praehistorica* XLIV: 78–86.

Núñez, M. 1990. On Subneolithic Pottery and its Adoption in Late Mesolithic Finland. *Fennoscandia archaeologica* VII: 27–52.

Nunez, M. 1998. Slates, the "Plastics" of Stone Age Finland. In L. Holm & K. Knutsson (eds.) *Proceedings from the Third Flint Alternatives Conference at Uppsala, Sweden, October 18–20, 1996*: 105–124. Occasional Papers in Archaeology 16.

Pesonen, P. 2004. Neolithic Pots and Ceramics Chronology: AMS-datings of Middle and Late Neolithic Ceramics in Finland. In P. Uino (ed.) *Fenno-Ugri et Slavi 2002: Dating and Chronology*: 87–97. Museoviraston arkeologian osaston julkaisuja N:o 10.

Pesonen, P. 2013a. Stratigraphy and Chronology of a Stone Age Pithouse in Raahe, Northern Ostrobothnia. In K. Johanson & M. Tõrv (eds.) *Man, His Time, Artefacts, and Places: Collection of Articles Dedicated to Richard Indreko / Inimene, tema aeg, esemed ja paigad: Richard Indrekole pühendatud artiklite kogumik*: 517–545. Muinasaja Teadus 19.

Rankama, T., Manninen, M. A., Hertell, E. & Tallavaara, M. 2006. Simple Production and Social Strategies: Do They Meet? Social Dimensions in Eastern Fennoscandian Quartz Technologies. In J. Apel & K. Knutsson (eds.) *Skilled Production and Social Reproduction. Aspects of Traditional Stone-Tool technologies. Proceedings of Symposium in Uppsala, August 20–24, 2003*: 245–261. Societas Archaeologica Upsaliensis Stone Studies 2.

Reimer, P., Austin, W., Bard, E., Bayliss, A., Blackwell, P., Bronk Ramsey, C., Butzin, M., Cheng, H., Edwards, R., Friedrich, M., Grootes, P., Guilderson, T., Hajdas, I., Heaton, T., Hogg, A., Hughen, K., Kromer, B., Manning, S., Muscheler, R., Palmer, J., Pearson, C., van der Plicht, J., Reimer, R., Richards, D., Scott, E., Southon, J., Turney, C., Wacker, L., Adolphi, F., Büntgen, U., Capano, M., Fahrni, S., Fogtmann-Schulz, A., Friedrich, R., Köhler, P., Kudsk, S., Miyake, F., Olsen, J., Reinig, F., Sakamoto, M., Sookdeo, A. & Talamo, S. 2020. The IntCal20 Northern Hemisphere Radiocarbon Age Calibration Curve (0–55 cal kBP). *Radiocarbon* 62 (4): 725–757.

Sarala, P. (ed.) 2012. *Lapin geologiset luonnonvarat 2010*. Acta Lapponica Fenniae 25.

Silvennoinen, A. 1998. Pohjois-Suomen liuskealueet, kerrosintruusiot ja granuliittialue. In M. Lehtinen, P. Nurmi & T. Rämö (eds.) *Suomen kallioperä: 3000 vuosimiljoonaa*: 141–163. Helsinki: Suomen Geologinen Seura.

Torvinen, M. 2000. Säräisniemi 1 Ware. *Fennoscandia archaeologica* XVI: 3–35.

Vuorinen, J. 1982. *Piikivi ja Suomen kampakeraaminen piikauppa*. Helsingin yliopiston arkeologian laitos, moniste n:o 30.

Notes

1 5471–4949 calBC (Hel-3666, 6250±100 BP) and 6233–5779 calBC (Hel-3669, 7140±110 BP) (Raike 1994).

2 Other knapped lithics include porphyry rocks (several in Kauniinmetsänniity 1), as well as individual pieces of sandstone, imported chert (Kivimaa, Kuuselankangas, Korvala), lydite (Latokangas) and jasper (Tervaniemi, Ahola).

3 In addition to bipolar and platform cores, Appendix 4 includes so-called irregular cores, undefined (multi-technology), fragmentary or exhausted cores, the proportion of which in assemblages ranges between 0 and 37%.

Appendix 1

Material used in the study. Phases: 1 – 6000–5000 calBC; 2 – 5000–4000 calBC; 3 – 4000–3500 calBC; 4 – 3500–3000 calBC; 5 – 3000–2000 calBC. Ceramics: Sär1 – Säräisniemi 1 Ware; Spr1 – Sperrings 1 Ware (also Older Early Comb Ware); Spr2 – Sperrings 2 Ware (also Younger Early Comb Ware); EAP – early asbestos-tempered pottery; TCW – Typical Comb Ware; SDCW – sparsely decorated Comb Ware; Kierikki – Kierikki Ware; Pöljä – Pöljä Ware. Excavations: researcher and year of investigation, the excavated area(s) included in the study are indicated in parentheses (research reports are archived by the Finnish Heritage Agency, Helsinki). Area: acreage of the excavated area included in the study (m²). KM: the studied collection numbers (the National Museum of Finland), sub-numbers are given when only a part of the material catalogued under this main number is included in the study.

No	Site	Municipality	Phase	Ceramics	Excavations (analysed areas)	Area (m²)	KM	Site type
1	Tervaniemi, Ahola	Taivalkoski	1	–	Saukkonen 1993 (areas 1 and 2); Raike 1994 (area 1, trenches 110 and 115): Raike 1995 (area 1995; x 102–104/y 534–536)	313	28128: 1–1302; 28687: 1–554; 28899: 1–815	Settlement
2	Vepsänkangas	Oulu	1+2	Sär1	Koivisto 1997 (all)	210	30561: 1–1001	Settlement
3	Veskankangas	Ii	1+2	–	Wallenius 1988 (area 1); 1989 (area 2); 1992 (part of area 6)	453	24423: 263–621; 24928: 1–1525; 27365: 1–1000	Settlement
4	Tainiaro	Simo	2	Spr1; Sär1; Spr2	Wallenius 1984 (x 509–532/y 989–1003); 1989 (x 510–518/y 984–992); 1990 (x 499–509/y 990–1003)	273	22398: 1–634; 24925: 734–1544; 25797: 762–2376	Settlement; burials
5	Latokangas	Oulu	2+3	Sär1, Spr1, EAP	Sarkkinen 1988 (main areas); Mäkivuoti & Sarkkinen 1989 (areas F, G, M); Mäkivuoti 1990 (areas 'Heimo' and 'Sär1')	740	24377: 1–374; 24740: 1–1159; 25731: 1–401	Settlement
6	Kivimaa	Kalajoki	3	TCW	Kotivuori & Halinen 1986 (areas I–III); Halinen 1987 (areas I–V)	242	23381: 1–560; 24123: 1–1192	Settlement
7	Kierikin sorakuoppa	Oulu	3	TCW	Schulz, E. 1986 (main area); Pärssinen 1987 (main area)	332	23431: 1–893; 23728: 1–768	Settlement; housepits
8	Kierikinkangas (north)	Oulu	3	TCW	Pesonen 1999 (sub-areas B and C, excavation areas 7–9)	147	31829: 2418–3116	Settlement
9	Kauniinmetsänniitty 1	Raahe	3	TCW	Pesonen 2007 (areas A–C; K, x 494–505/ y 702–716)	148	36937 (1938 sub-numbers)	Settlement; housepits
10	Kuuselankangas	Oulu	3+4	TCW; Kierikki; SDCW	Koivunen & Korolainen 1995 (main area); Koivunen & Franzen 1996 (main area); Halinen 1996 (main areas); Pesonen 1999 (area 1)	529	30666: 1023–2676; 30667: 20–759; 29907: 1–2085; 32220: 1–7243	Settlement; housepits
11	Purkajasuo-Korvala	Oulu	4	Pöljä	Schulz, H.-P. 2000 (area 9)	176	32134: 1–2267	Settlement; housepits
12	Hangaskangas E	Oulu	5	–	Pesonen 2012 (areas 7a; 7b; 9)	48	39158: 1147–1312, 1313–1327, 1439–1756	Settlement

Appendix 2

Radiocarbon dates of the studied materials. Dates of loose charcoals from the cultural layer and dates with standard deviation greater than ±75 years have been excluded. All dates are calibrated with OxCal 4.4 (Bronk Ramsey 2009) using IntCal20 (Reimer et al. 2020) and given with a 95.4% probability. For abbreviations of pottery types, see Appendix 1.

No	Site	Lab-code	BP	±	calBC	(2 σ)	$\delta^{13}C$	Material (type/context)	Pottery type	Reference
1	Tervaniemi, Ahola	–	–	–	–	–	–	–	–	–
2	Vepsänkangas	GrA-63484	6135	40	5211	4952	-26,58	crust	Sär1	Nordqvist & Mökkönen 2016
		Hela-236	6120	75	5290	4844	-26,30	crust	Sär1	Koivisto 1997; Torvinen 2000
		Hela-235	6065	75	5209	4797	-27,50	birch bark tar ('chewing resin')	–	Koivisto 1997; Torvinen 2000
		Hela-128	5995	65	5042	4721	-22,20	crust	Sär1	Torvinen 2000
		Hela-312	5990	60	5027	4723	-27,30	birch bark tar ('chewing resin')	–	Koivisto 1998
3	Veskankangas	GrA-63889	5980	40	4991	4731	-27,90	burnt bone (elk)	–	This work
4	Tainiaro	GrA-63478	5615	40	4537	4356	-25,56	crust	Spr2	Nordqvist & Mökkönen 2016
		GrA-63480	5735	40	4691	4459	-26,42	crust	Sär1	Nordqvist & Mökkönen 2016
		GrA-63483	5775	40	4721	4504	-27,79	crust	Spr1	Nordqvist & Mökkönen 2016
5	Latokangas	GrA-63485	6010	40	5002	4793	-24,91	crust	Sär1	Nordqvist & Mökkönen 2016
		GrA-63486	5025	35	3946	3711	-27,89	crust	Sär1	Nordqvist & Mökkönen 2016
6	Kivimaa	GrA-63516	4905	35	3769	3637	-26,75	crust	TCW	Nordqvist 2018
		GrA-63517	4765	45	3641	3378	-26,08	crust	TCW	Nordqvist 2018
7	Kierikin sorakuoppa	GrA-63488	4850	35	3707	3528	-25,54	crust	TCW	Mökkönen & Nordqvist 2018
		GrA-63487	4790	35	3645	3386	-27,36	birch bark tar on pottery	TWC	Mökkönen & Nordqvist 2018

No	Site	Lab-code	BP	±	calBC	(2 σ)	δ¹³C	Material (type/context)	Pottery type	Reference
8	Kierikinkangas (north)	Hela-408	4820	65	3759	3378	-27,20	birch bark tar on pottery	TCW	Pesonen 2004
9	Kauniinmetsänniitty 1	GrA-63544	5055	35	3959	3716	-25,86	crust	TWC	Nordqvist 2018
		GrA-63542	4935	35	3781	3644	-28,49	crust	TWC	Nordqvist 2018
		GrA-63546	4895	35	3769	3633	-26,10	crust	TWC	Nordqvist 2018
		Hela-1712	4770	40	3641	3381	-29,60	birch bark tar on pottery	TWC	Pesonen 2013a
		Hela-1713	4730	40	3632	3376	-25,40	crust	TWC	Pesonen 2013a
		Hela-1714	4690	40	3624	3368	-19,60	burnt bone (seal)	–	Pesonen 2013a
		Hela-1711	3935	35	2566	2298	-27.50	birch bark tar ('chewing resin')	–	Pesonen 2013a
10	Kuuselankangas	GrA-63491	4990	35	3940	3652	-25,91	crust	TCW	Mökkönen & Nordqvist 2018
		Hela-464	4625	70	3627	3103	-25,30	charcoal (posthole)	–	Pesonen 2000
11	Purkajasuo-Korvala	GrA-63504	4535	35	3366	3101	-27,08	birch bark (housepit, structure)	–	Mökkönen & Nordqvist 2018
		Hela-136	4475	60	3360	2935	-28,80	crust	Pöljä	Pesonen 2004
		GrA-63505	4455	35	3340	2937	-27,71	crust	Pöljä	Mökkönen & Nordqvist 2018
12	Hangaskangas E	Ua-45449	3775	40	2341	2038	-28,40	bone (seal)	–	Pesonen 2013b
		Ua-454450	3695	35	2200	1972	-27,70	bone (seal)	–	Pesonen 2013b

Appendix 3

Volume of the studied lithics per site. HQQ –high-quality quartz, i.e. microcrystalline variants of quartz (different colour variants of rock crystals). For abbreviations of pottery types, see Appendix 1.

Site	Ceramics	Phase	Quartz (g/pcs)	Quartzite (g/pcs)	Slate (g/pcs)	Flint (g/pcs)	Other (g/pcs)	All (g/pcs)	High-quality quartz (% of cataloque nos)
Tervaniemi, Ahola	–	1	5528/3111	162/14	128/58	–	93/15	5911/3198	6
Vepsänkangas	Sär1	1+2	8944/1656	803/86	892/185	3/3	–	10 641/1930	2
Veskankangas	–	1+2	41 008/8961	94/22	18 237/2644	8/2	–	59 346/11 629	3
Tainiaro	Spr1; Sär1; Spr2	2	19 066/3030	177/19	114 125/11 523	97/7	98/7	133 562/14 586	2
Latokangas	Sär1; Spr1; EAP	2+3	14 755/3537	1899/178	1481/347	<1/1	9/2	18 144/4065	3
Kivimaa	TCW	3	33 834/7209	99/7	147/29	6/5	45/7	34 131/7257	10
Kierikin sorakuoppa	TCW	3	12 633/1963	1185/15	2957/177	106/81	17/3	16 909/2240	8
Kierikinkangas (north)	TCW	3	2533/1163	43/12	351/31	9/1	–	2936/1212	1
Kauniinmetsänniitty 1	TCW	3	5869/1665	71/3	1009/113	10/4	597/58	7555/1843	–
Kuuselankangas	TCW; Kierikki; SDCW	3+4	41 700/21 172	1298/78	6758/570	32/15	235/14	50 023/21 849	1
Purkajasuo-Korvala	Pöljä	4	18 668/12 172	620/77	2903/205	5/6	12/7	22 209/12 467	–
Hangaskangas E	–	5	1113/733	139/7	826/360	–	–	2078/1100	–
								363 445/83 376	

Appendix 4

Basic metrics for the studied quartz lithics. For abbreviations of pottery types, see Appendix 1.

Site	Ceramics	Phase	Flakes (pcs)	Tools (pcs)	Tool-to-flake ratio (pcs)	Bipolar cores (pcs)	Platform cores (pcs)	Irregular cores (pcs)	Bipolar-to-platform core ratio	Cortex (% of catalogue nos)
Tervaniemi, Ahola	–	1	2931	181	6,2	55	4	6	13,8	5
Vepsänkangas	Sär1	1+2	1602	54	3,4	18	–	4	–	11
Veskankangas	–	1+2	8624	321	3,7	73	9	24	8,1	14
Tainiaro	Spr1; Sär1; Spr2	2	2961	69	2,3	36	1	4	36,0	18
Latokangas	Sär1; Spr1; EAP	2+3	3444	93	2,7	52	8	5	6,5	5
Kivimaa	TCW	3	7024	284	4,0	47	10	34	4,7	18
Kierikin sorakuoppa	TCW	3	1844	119	6,5	21	16	10	1,3	8
Kierikinkangas (north)	TCW	3	1116	47	4,2	–	3	–	–	5
Kauniinmetsänniitty 1	TCW	3	1625	40	2,5	4	–	–	–	6
Kuuselankangas	TCW; Kierikki; SDCW	3+4	20 849	323	1,5	110	21	26	5,2	6
Purkajasuo-Korvala	Pöljä	4	12 011	161	1,3	36	16	9	2,3	9
Hangaskangas E	–	5	732	1	0,1	–	–	–	–	17

Stone Age Settlements in the Pskov Region (Northwest Russia): A Review of Old Materials and New Fieldwork

Evgeniia S. Tkach

Evgeniia S. Tkach, Institute for the History of Material Culture, Russian Academy of Sciences, Dvortsovaya emb., 18, 191186 Saint-Petersburg, Russia, e-mail: evgeniia.tkach@gmail.com

Abstract

The paper presents an overview of the Stone Age materials discovered in the central and northern parts of the Pskov district (Northwest Russia). There are 15 sites where stone and ceramic materials were identified. Some materials from the sites were lost, while others were identified only by artefacts from quartz, flint or ceramic fragments, which renders their cultural identification questionable. The most representative collection from the territory discussed was obtainned from site 10 near the lake of Belaya Struga with stone (flint, slate, quartz) and ceramic artefacts. Analogies for pottery could be traced among Typical Comb Ware materials in Latvia, Estonia and the nearest Russian territories (Leningrad district), dating to the 5th millennium-4th millennium BC. Some micro-blades and micro-scrapers were also discovered, which could perhaps be attributed to the Mesolithic period. Further studies in this region seem promising in terms of searching for the cultural layer "in situ" of both the Mesolithic and Neolithic periods.

1 Introduction

Generally, the Pskov region can be included in the "Circum-Baltic" space of Stone Age cultures, but this territory is still a "white spot" on the Stone Age map of north-western Russia. The sites in question (including pile-dwelling settlements) were only studied in the southern part of the region (Miklyaev 1969). Numerous sites have been identified in Estonia (Jaanits 1959; Johanson et al. 2014; Roio & Kriiska 2015; Sikk et al. 2020; etc.) and Latvia (Loze 1979; 2015; Macāne & Nordqvist 2021; Zagorskis 1963; 1987; etc.), and a lot of Stone Age sites are also present in the Leningrad region (Gerasimov et al. 2010; Kriiska & Gerasimov 2014; etc.).

The Closest sites in Russia are located at the western part of the Leningrad district – Sokolok and Syaberskoe III (Timofeev 1985; 1993). Outside of Russia, the nearest sites are located in south-eastern Estonia – Tamula I (Indreko 1945; Jaanits 1984; Jaanits et al. 1982), and in eastern Latvia – Osa, Piestina, Abora I, Asne I and others of the Lubāna Lowland (Loze 1979; 2000; Zagorskis 1965; Zagorskis et al. 1984).

When it comes to the central and northern parts of the Pskov region, however, only sporadic sites are known. Hence this region is of great interest for Stone Age archaeological surveys and excavations.

2 Stone Age sites in the Pskov region

The first mentions of Stone Age materials from the Pskov region date back to the beginning of the 20th century. They were signed on the archaeological map that was made for the XIV Russian Archaeological Congress in 1913 (Trudy 1914), but until the mid-20th century only a few sites with flint

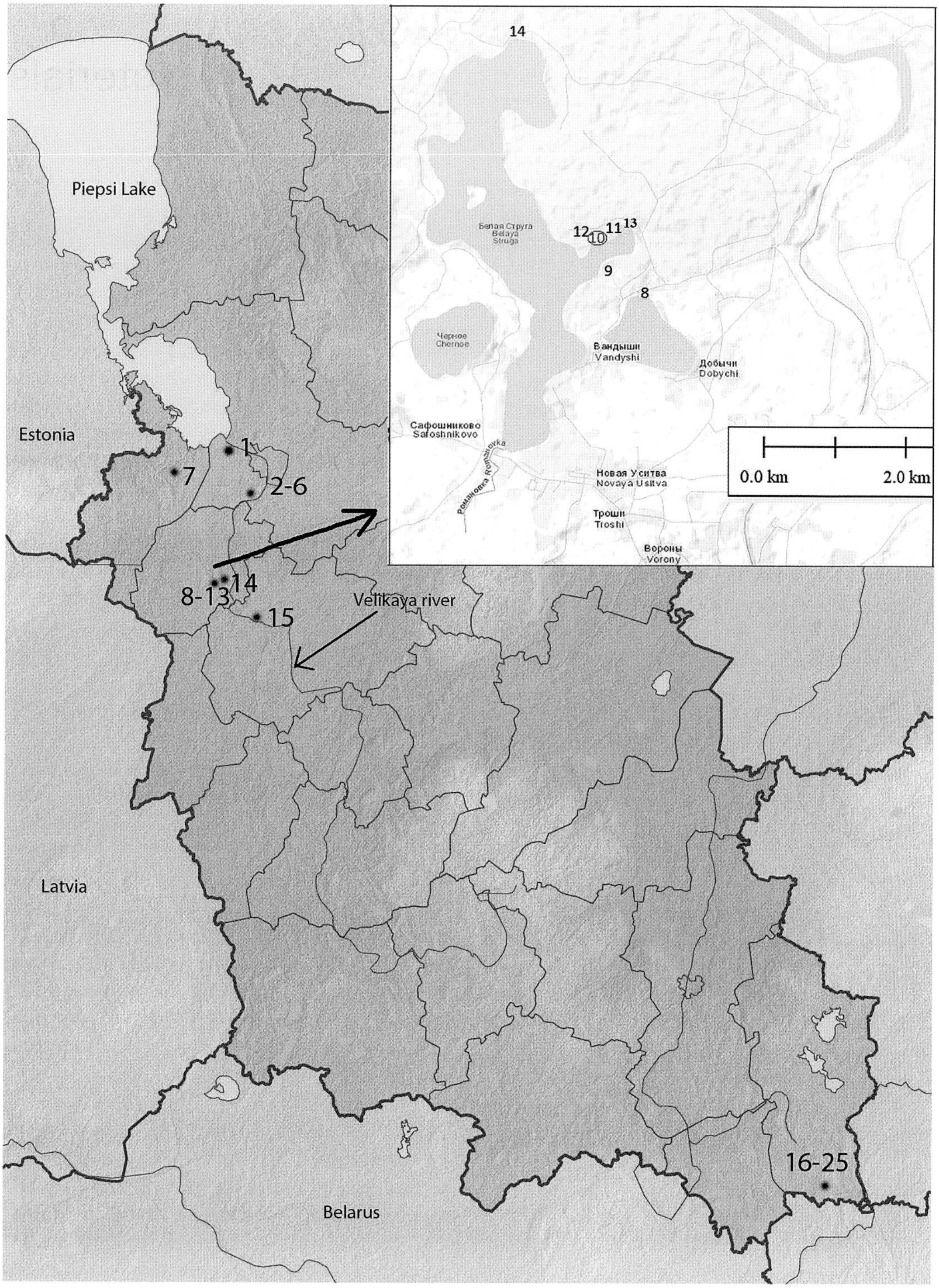

Figure 1. Locations with Stone Age materials in the Pskov region (1 – 15 – locations, found by N. N. Gurina and A. N. Scheglov; 16 – 25 – locations, found by A. M. Miklyaev).

and ceramic artefacts were known in this territory.

Further investigations in the region were connected with Nina N. Gurina's fieldworks in the 1950s. The main topic of her study was a history of the population of Northwest Russia in the Stone Age (Gurina 1955; Mazurkevich 2008). She surveyed the shores of Lake Peipus in the northern part of the Pskov region and a course of the Velikaya River from the town of Ostrow to the river mouth at Lake Peipus in its central part.

During the 1955–57 field seasons, Gurina uncovered 11 sites (Gurina 1955), but for now the only available tools for locating the aforementioned sites are short descriptions and small-scale sketches of artefact findspots with no exact geographical coordinates.

All the sites are located on the left bank of the Velikaya River (Fig. 1). One of these sites, – Gogolevka (Fig. 1: 3), near Pskov, was already known earlier, – as research there was conducted in 1912 by Alexander Mednis, the employer of the Pskov Archaeological Society. The cursory results were published by Sergey Parkhomenko in 1916 (Parkhomenko 1916: 6–7). The site was located on the surface of a destroyed dune. Fragments of ceramic vessels with horizontal lines of pits and comb stamps, pieces of burnt bones (unidentified) as well as flint artefacts (flakes, chips, fragments of knives and arrowheads, scrapers) were found in the dune's blows. Unfortunately, the area of this site was destroyed by modern activities and the archaeological collection lost.

Some of the other potsherds discovered (Fig. 1: 15) could be attributed to Asbestos Ware (close to the Orovnavolok–XVI type – see: Zhulnikov 1999: 50) or Comb Ware (Fig. 1: 5, 9), but the fragments are largely indefinable due to their size.

3 Belaya Struga sites

In 1957 Aleksander Shcheglov carried out an archaeological survey along the shores of Belaya Struga lake in the Palkino district, in the central part of the Pskov region (Shcheglov 1957). He revealed six sites (Fig. 1: 8–13):

1) Site 1 by A. N. Shcheglov (Fig. 1: 8) by the northern shore of Lake Shchadritskoe. Finds: flint chips and small fragments of ceramic vessels.
2) Site 2 by A. N. Shcheglov (Fig. 1: 9) by the southern cape of the eastern bay of Belaya Struga lake. Finds: Pit-Comb Ware fragments and a few flint chips.
3) Site 3 by A. N. Shcheglov (Fig. 1: 10–12) by the northern cape of the eastern bay of Belaya Struga lake, opposite site 2. A. N. Shcheglov has defined three concentrations of finds there: a) western, on the cape. Finds: numerous flint artefacts and several ceramic fragments (Fig. 1: 10); b) materials from the central part, including thin-walled ceramics with cut-in ornament of straight lines forming shaded triangles (Fig. 1: 11); c) eastern. Finds: flint artefacts including large scrapers, fragments of thick-walled ceramic vessels (Fig. 1:12).
4) Site 4 by A. N. Shcheglov (Fig. 1: 13) located 150–200 metres to the east from site 3 and containing two Mesolithic blades (according to his description).

The archaeological surveys at the shores of Belaya Struga lake were continued in 1958 by Nina Gurina, during which she found a site on the left bank of the River Struglitsa containing Mesolithic materials (according to her notes) (Fig. 1: 14).

In 2018–2019 the fieldworks were continued by the author, during which site 10 was revealed. A test pit of 2x2 sq.m. was excavated. The total area of the site is around 90 sq.m. The finds contained stone artefacts, pottery fragments and a large number of pieces of burnt bones. Unfortunately, these materials cannot be separated based on stratigraphy or spatial analysis. The materials were deposited at the edge of the terrace in the sandy soil. The excavation was conducted in 5–7 cm horizons and the finds were documented with a laser total station. Neolithic and Iron Age ceramics were found together, and pieces of burnt bones were found in all horizons. The cultural layers are not distinguished by lithology.

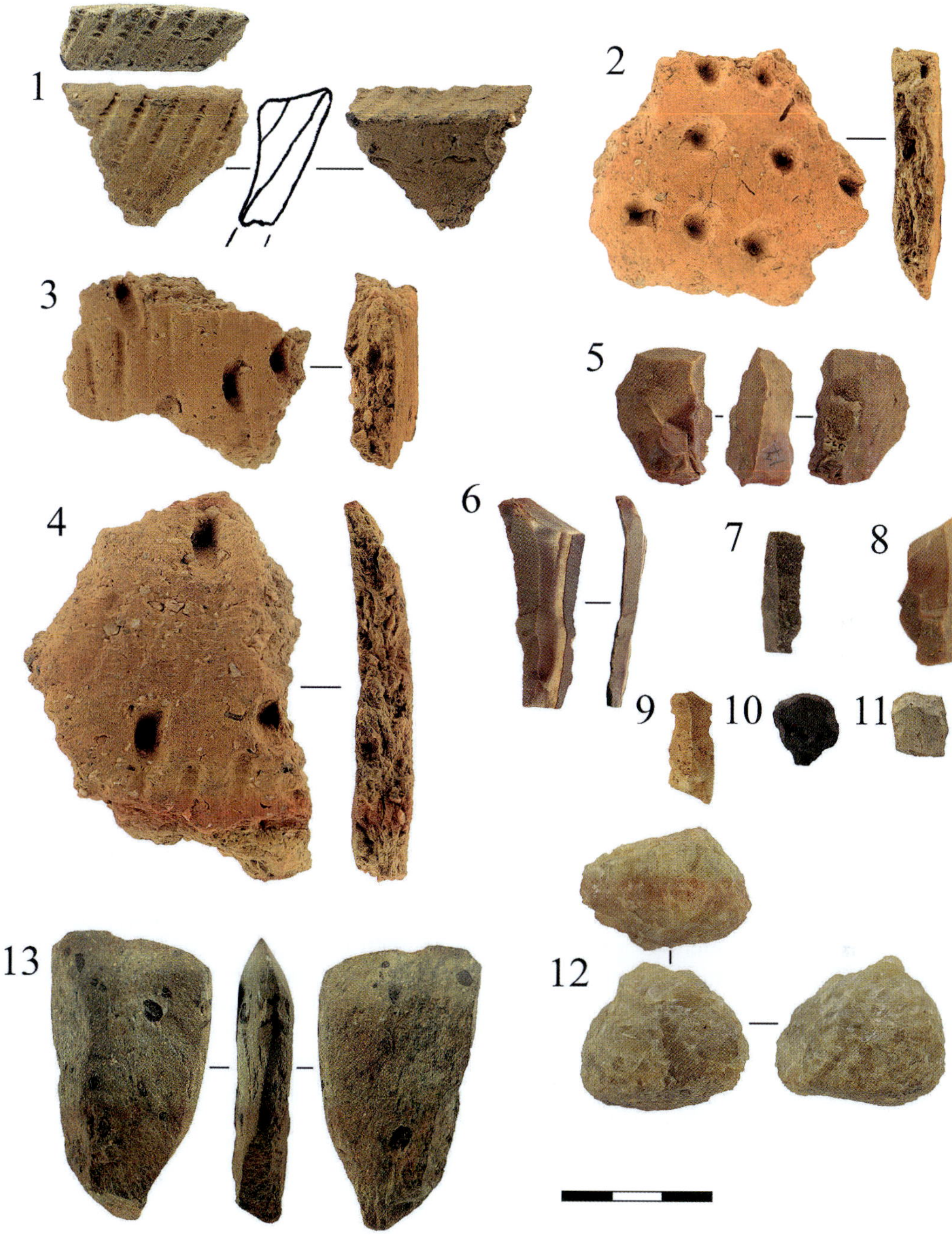

Figure 2. Materials from the Belaya Struga site, no. 3: 1 – 4 – ceramic; 5 – 11 – flint; 12 – quartz; 13 – slate.

Type	Material			
	flint	slate	quartz	
microscrapers	4			
microblades	2			
blades and fragments	2			
cores	2		1	
flake-blades	3			
flakes	2	2		
tools fragments with polishing		2		
axe		1		
chips	64	4	3	
total	**79**	**9**	**4**	**92**
total (%)	**86%**	**10%**	**4%**	**100%**

Table 1. Stone materials from site 10 (Belaya Struga lake).

So far, a total of five Stone Age sites have been identified along the shores of Belaya Struga lake. The most representative archaeological collection was obtained from site 10.

3.1 Stone artefacts

More than 103 fragments of stone artefacts were found, mostly grey and black flint. Some slate and quartz artefacts were also found (Table 1).

The materials consisted mostly of debris (86 %). The collection includes two microblades (Fig. 2: 7), four small scrapers on the ends of blades (Fig. 2: 10), two fragments of flint amorphic cores (Fig. 2: 5) and one quartz core (Fig. 2: 12). The slate materials include an axe (Fig. 2: 13), flakes, chips and fragments of slate tools with polishing.

3.2 Ceramic vessels

Overall, 196 fragments of ceramic vessels were found. Some of them were attributed to the Iron Age – such as a flat-bottomed vessel with an artificial gruss admixture. The rim is slightly bent outwards. The vessel has no ornamentation.

There are a minimum of seven vessels (according to a complex of features: moulding clay, the rim's profile, ornamentation elements and motifs, technology of making pottery) with pit and comb stamps in ornamentation (Fig. 3). They have been made via the patching technique, the thickness of the wall ranges from 6 to 9 mm, and the surface is well-smoothed. The clay mass contains an artificial gruss admixture. The rims are straight and thickened. The ornamentation has been made with deep conical pits (Fig. 2: 2), less often in combination with a comb stamp (Fig. 2: 4). Some of the rims have only been ornamented with comb stamps. Sometimes the edge of the vessel is also ornamented (Fig. 2: 1).

The ceramic collection also features fragments of a vessel with organic temper as an admixture. The wall's thickness is 5 mm. The rim is bent outwards. The ornamentation features a horizontal line of the notches' imprints.

4 Conclusions

There are 15 sites containing Stone Age materials in the central and northern parts of the Pskov region. Most of them are sites that were uncovered in the middle of the 20th cen-

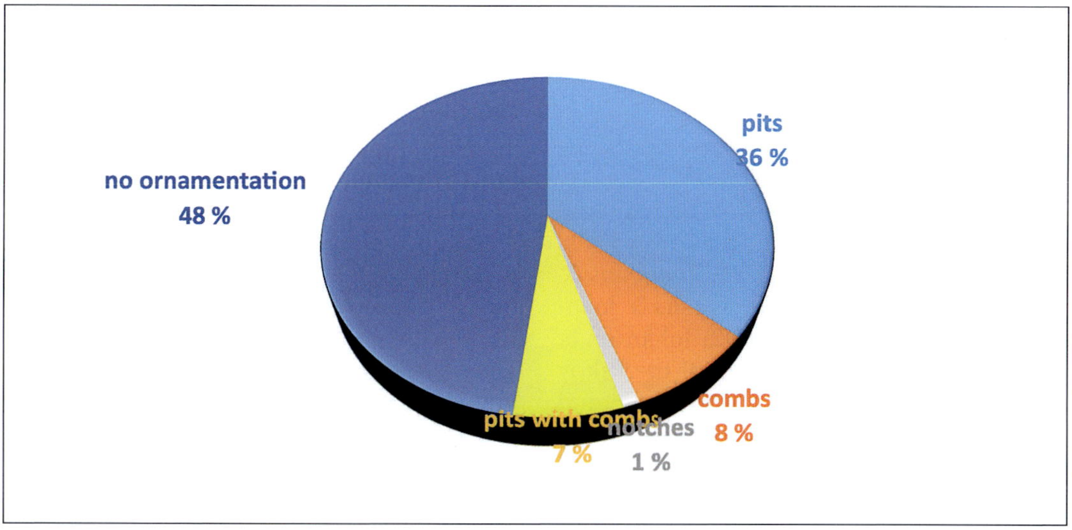

Figure 3. Ornamentation stamps and their combination on ceramics (Belaya Struga lake, site 10).

tury by Nina Gurina. Some materials were lost (like at Gogolevka), while others only consist of artefacts from quartz, flint, or ceramic fragments.

The most representative collection from the territory discussed was obtained from site 10 near Belaya Struga lake. This is the multilayer site. The cultural layer consists of stone, ceramic materials and pieces of burnt bones. Some pottery (flat-bottomed vessel with artificial gruss admixture without ornamentation) has analogies in Iron Age materials.

The ornamentation of ceramic fragments with pits and combs could be attributed as Neolithic. There are a lot of similar ceramics among the materials of the Typical Comb Ware, which was used in north-eastern Europe in the 4th millennium-5th millennium BC (Nordqvist & Kriiska 2015). Sites with Typical Comb Ware materials have been identified in Latvia, Estonia, Finland and Russia (the Karelian Isthmus, Leningrad district, etc.).

The nearest sites with Comb Ceramic pottery (Loze 1984) are located in Latvia (site Zvidze – see: Loze 1988: 268; site Sulka – see: Loze 1988: 268), Estonia (burial Tamula I – see: Indreko 1945; Jaanits 1984) and the Leningrad district. The Comb Ceramic pottery from the Syaberskoe III site in Russia dates to the end of the 3rd millennium BC (Timofeev et al. 2004: 96–97). Perhaps the Gogolevka site (according to the description) could also be attributed to this cultural group. The nearest Typical and Late Comb Ware materials in the burial Tamula I in Estonia date to the end of the 5th millennium-middle of the 4th millennium BC (4200–3495 cal BC) (Kriiska et al. 2007: 107), while in Latvia the Comb Ware materials from the site of Zvidze date to the first half of the 3rd millennium BC (Loze et al. 1984: 54).

Cultural identification of the stone (flint) artefacts from site 10 is problematic. There are no tools which could be attributed to some cultural tradition or archaeological culture. Most of the materials in the collection consist of debris. The slate axe has analogies among the Comb Ware materials (Loze 1988: tab. XXXII). Quartz tools also feature among the artefacts of the Comb Ware culture in Latvia and Russia. This material was also used during the Mesolithic in Estonia (Rostedt & Kriiska 2019), but it was widespread during the Neolithic.

The analysed collection contains some micro-blades and micro-scrapers that have been made from the local materials. These materials are not characteristic of the Middle Neolithic, but are representative of the Mesolithic. According to Gurina's description, site 14 at the River Struglitsa can be attributed to the Mesolithic.

A lot of Mesolithic sites have been discovered in Latvia (Zvejnieki II, the lower layer, Suliagals, Osa, etc.) (Loze 1988; Zagorskis & Zagorskis 1977) and Estonia (Jaanits & Jaanits 1978: 63; Jaanits et al. 1982; Johanson et al. 2021). In the southern part of the Pskov region materials of the final Paleolithic and the Early Mesolithic (Ivantsov Bor, Baburova Gora, Lukashenki I and III) were also found (Lisitsyn 2003: 40– 41; Zheltova 2003: 39).

The late stage of the Mesolithic (end of the 8th millennium – middle of the 5th millennium BC) at the eastern part of the Baltic coast (Gerasimov & Kriiska 2018: 307; Kriiska & Gerasimov 2014) is marked by the use of local materials, small sized tools (micro-scrapers) and the absence of morphologically expressed tool types (Kriiska & Gerasimov 2014: 18).

Thus, we can suppose that more Mesolithic sites could be discovered near the lake of Belaya Struga. And perhaps the presence of microblades and microscrapers at site 10 is related to this period.

We hope that during the future excavations at the central part of the Pskov region more cultural and chronological data will be obtained.

Acknowledgements

The study was carried out within the framework No. 0184–2019–0002.

The author is very grateful to Dmitriy V. Gerasimov and the reviewers for their comments on this article.

References

Archival sources

Gurina 1955 = Гурина Н.Н. *Отчет о работе отряда по обследованию Псковской, Великолукской, Калининской обл.* Архив ИИМК РАН. Ф. 35, оп. 1955. 97 л.

Shcheglov 1957 = Щеглов А.Н. *Стоянки озера Белая Струга.* Архив ПГОИАХМЗ. Оп. 2. 34485/2. 7 л.

Literature

Gerasimov, D. & Kriiska A. 2018. Early–Middle Holocene archaeological periodization and environmental changes in the Eastern Gulf of Finland: Interpretative correlation. *Quaternary International* 465: 298–313.

Gerasimov, D., Kriiska, A. & Lisitsyn, S. 2010. Colonization of the Gulf of Finland (the Baltic Sea) Coastal Zone in the Stone Age. In A. V. Golovnev (ed.) *III Northern Archaeological Congress*: 29–54.

Indreko, R. 1945. Märkmeid Tamula leiu kohta. *Suomen Muinaismuistoyhdistyksen Aikakauskirja* XLV: 26–43.

Jaanits, L. 1984. Die kennzeichende Züge der Siedlung Tamula. *ISKOS* 4: 183–193.

Jaanits 1959 = Янитс Л. Ю. *Поселения эпохи неолита и раннего металла в приустье р. Эмайыги (Эстонская ССР).* Талин: Академия наук Эстонской ССР, Институт истории.

Jaanits, L. & Jaanits, K. 1978. Ausgrabunden der Frühmesolitischen Siedlung von Pulli. *Известия АН Эстонской ССР, общественные науки* 27 (1): 56–63.

Jaanits, L., Laul, S., Lõugas, V. & Tõnisson, E. 1982. *Eesti esiajalugu*. Tallinn: Eesti Raamat.

Johanson, K., Kadakas, U., Tõrv, M. & Vindi, A. 2014. Excavations on the island of Kloostrisaar. Some reflections on the Stone Age settlement pattern in South Estonia. *Archaeological fieldwork in Estonia 2013*: 33–44.

Johanson, K., Kriiska, A., Aruväli, J., Somelar, P., Sikk, K. & Sepp, L. 2021. Local or imported? Tracking the provenance of flint raw materials of the Mesolithic habitants of Estonia and northern Latvia with the help of geochemical methods. In: D.Borić, D.Antonović, B. Mihailović (eds.). *Foraging Assemblages* 1: 123–128.

Kriiska & Gerasimov 2014 = Крийска, А., Герасимов, Д. В. 2014. Период позднего мезолита в восточной части Балтийского моря: формирование берегового расселения от Рижского залива до Выборгского. In В. Н. Карманов (ed.) *От Балтики до Урала: изыскания по археологии каменного века*: 5–36.

Kriiska, A., Lõugas, L., Lõhmus, M., Mannermaa, K. & Johanson, K. 2007. New AMS dates from Estonian Stone Age burial sites. *Estonian Journal of Archaeology* 11 (2): 83–121.

Lisitsyn 2003 = Лисицын, С. Н. 2003. Проблема разделения смешанных комплексов финального палеолита и мезолита Двинско-Ловатского междуречья. In Г. В. Виллинбахов, Б. С. Короткевич, А. Н. Мазуркевич

& Ю. Ю. Пиотровский (eds.) *Древности Подвинья: исторический аспект*: 40–47.

Loze 1979 = Лозе, И. А. 1979. *Поздний неолит и ранняя бронза Лубанской равнины*. Рига: Зинанте.

Loze 1984 = Лозе, И. А. 1984. Памятники гребенчато-ямочной керамики в Восточной Прибалтике. In В. А. Рыбаков (ed.) *Новое в археологии СССР и Финляндии*: 28–33.

Loze 1988 = Лозе, И. А. 1988. *Поселение каменного века Лубанской низины: Мезолит, ранний и средний неолит*. Рига: Зинанте.

Loze, I. 2000. The early Neolithic at the Ica settlement site (Lake Lubāna Depression). *Lietuvos Archeologija* 19: 203–220.

Loze, I. 2015. *Lubāna mitrāja apdzīvotība akmens laikmetā*. Rēzekne: Rēzeknes Augstskola

Loze et al. 1984 = Лозе, И. А., Лийва, А., Стелле, В. Я., Эберхардс, Г.Я., Якубовская, И. Я. Звизде – многослойное поселение эпох мезолита и неолита на Лубанской низине (Латвийская ССР). In А. А. Величко, Л. В. Кольцов, Н. А. Хотинский & Т. А. Серебрянная (eds.) *Археология и палеогеография мезолита и неолита Русской равнины*: 40–55.

Macāne, A. & Nordqvist, K. 2021. More Than Just Zvejnieki: An Overview of Latvian Stone Age Burials. *European Journal of Archaeology* 24 (3): 299–323. doi:10.1017/eaa.2020.64

Mazurkevich 2008 = Мазуркевич, А. Н. 2008. Находки каменного века с северных территорий Псковской области. In И. К. Лабутина (ed.) *Археология и история Пскова и Псковской земли*: 186–194.

Miklyaev 1969 = Микляев, А. М. 1969. Памятники Усвятского микрорегиона. Псковская область. *Археологический сборник Государственного Эрмитажа* 11: 18–40.

Nordqvist, K. & Kriiska, A. 2015. Towards Neolithisation: The Mesolithic-Neolithic transition in the central area of the eastern part of the Baltic Sea. In J. Kabaciński, S. Hartz, D. Raemaekers & T. Terberger (eds.) *The Dąbki Site in Pomerania and the Neolithisation of the North European Lowlands (c. 5000–3000 calBC)*: 537–556.

Parkhomenko 1916 = Пархоменко, С. 1916. К вопросу о каменном веке в Псковской губернии. *Труды Псковского Археологического Общества 1915–1916 гг.* 12: 3–9.

Roio, M. & Kriiska, A. 2015. Archaeological survey in the northern and north-western parts of Lake Peipsi. *Archaeological fieldwork in Estonia 2015*: 225–234.

Rostedt, T. & Kriiska, A. 2019. Quartz Assemblage from the Early Mesolithic Helvetinhaudanpuro Settlement Site in Eastern Central Finland. *Vestnik of Saint Petersburg University. History* 64 (2): 639–665. https://doi.org/10.21638/11701/spbu02.2019.213

Sikk, K., Kriiska A., Johanson, K., Sander, K. & Vindi, A. 2020. Environmental and settlement location choice in Stone Age Estonia. *Estonian Journal of Archaeology* 2020, 24, 2: 89–140. https://doi.org/10.3176/arch.2020.2.01

Timofeev 1985 = Тимофеев, В. И. 1985. Работы по изучению памятников Ленинградской области в 1982–1984 гг. In В. П. Шилов & Дж. А. Халилов (eds.) *Всесоюзная археологическая конференция. Достижения советской археологии в XI пятилетке*: 346–348.

Timofeev 1993 = Тимофеев, В. И. 1993. Памятники мезолита и неолита региона Петербурга и их место в системе культур каменного века Балтийского региона. In В. М. Массон, Е. Н. Носов & Е. А. Рябинин (eds.) *Древности Северо-Запада*: 8–34.

Timofeev et al. 2004 = Тимофеев, В. И., Зайцева, Г. И., Долуханов, П. М. & Шукуров А. М. 2004. *Радиоуглеродная хронология неолита Северной Евразии*. СПб: Теза.

Trudy ... 1914 = *Труды Псковского Археологического Общества 1913–1914 гг.* 10. Посвящается XVI Всероссийскому Археологическому Съезду. Псков: Электрич. типо-лит. Псков. Губерн. Земства.

Zagorskis, F. 1963. Kreiču neolīta apmetne. *Latvijas PSR Zinātņu Akadēmijas Vēstis* 4: 23–35

Zagorskis, F. 1965. Jauni materiāli par neolitu Latvijas austrumu dalā. *Latvijas PSR zinātņu akadēmijas vēstis* 6: 35 – 50.

Zagorskis, F. 1987. *Zvejnieku akmens laikmeta kapulauks*. Rīga: Zinātne.

Zagorskis & Zagorskis 1977 = Загорскис, Ф. & Загорская, И. 1977. Мезолит Латвии. *Краткие сообщения Института археологии* 149: 69–75.

Zagorskis et al. 1984. = Загорскис, Ф. А., Эберхардс, Г. Я., Стелле, В. Я., Якубовская, И. Я. 1984. Оса – многослойное поселение эпох мезолита и неолита на Лубанской равнине (Латвийская ССР). In А. А. Величко, Л. В. Кольцов, Н. А. Хотинский & Т. А. Серебрянная (eds.) *Археология и палеогеография мезолита и неолита Русской равнины*: 55–67.

Zheltova 2003 = Желтова, М. Н. 2003. Кремневые индустрии памятников финального палеолита и мезолита бассейна реки Усвячи. In Г. В. Виллинбахов, Б. С. Короткевич, А. Н. Мазуркевич & Ю. Ю. Пиотровский (eds.) *Древности Подвинья: исторический аспект*: 27–39.

Zhulnikov 1999 = Жульников, А. М. 1999. Энеолит Карелии (памятники с пористой и асбестовой керамикой). Петрозаводск: ИЯЛИ КарНЦ.

Abbreviations

ИЯЛИ – Институт языка и литературы
КарНЦ – Карельский научный центр
ПГОИАХМЗ – Псковский государственный объединенный историко-архитектурный и художественный музей-заповедник

Excavations of Bronze Age Burial Cairns near the Village of Bolshoy Bor in 2017 (Northern Coast of the Gulf of Finland)

Andrey Gorodilov

Andrey Gorodilov, Institute for the History of Material Culture, Russian Academy of Sciences, Dvortsovaya emb. 18, 191186 St. Petersburg, Russia, e-mail: ramapitec@yandex.ru

Abstract

In 2017, an expedition of History of Material Culture (IIMK) of the Russian Academy of Sciences (RAS) investigated a group of stone cairns situated near the village of Bolshoy Bor on the northern coast of the Gulf of Finland. The group in question consisted of four mounds located on a granite rock corresponding to the ancient seashore. In total, three cairns have been investigated, as along with the stone structures surrounding them. Over the course of the investigations, it was revealed that the objects were stone cairns dating to the Bronze Age. Inside the cairns, the remains of bonfires, accumulations of calcined bones and accompanying grave goods (a bronze knife) were uncovered. These findings suggest that the site was a funerary and ritual complex. In addition, elements of stone masonry were found in the cairns, viz. stone crepis walls constructed of massive boulders, circular structures, etc.

1 Introduction

The northern coast of the Gulf of Finland east of the Russian-Finnish border and the coastal regions of Finland are geomorphologically the same space. However, in terms of the archaeological study of antiquities from the Bronze Age, these territories differ from each other quite strongly. Until now, the monuments of this period on Russian territory have practically been left unstudied due to several factors, including the inaccessibility of the border area for examination until the end of the 1990s. In 2017, a group of Bronze Age burial cairns located to the west of Vyborg near Bolshoy Bor (Jokikylä, Säkkijärvi district) was excavated, with the aim of clarifying the cultural and chronological attribution of this group, as well as tracing the local features of the funeral rite in the eastern part of the area of the Northern European culture of the Bronze Age.

2 Information gained during the pre-War period on Bronze Age sites in the Gulf of Finland region

The western area of the Karelian Isthmus and the Outer Islands (Suomenlahden ulkosaaret) archipelago of the Gulf of Finland were part of Finland before the Second World War. During this time, a significant quantity of data on the archaeological sites of this region, including Bronze Age antiquities, were accumulated (Fig. 1). This information is summarised in works by T. Miettinen, P. Uino, A. I. Saksa and M. Lavento (Lavento 2001: 244–256; 2003; Miettinen 1996; Saksa 2014; Uino 1997: 104–108).

The maritime zone of Bronze Age southern and south-western Finland is part of the area of the distribution of the Nordic Bronze Age culture, with its main seat located in southern Scandinavia. Stone cairns, or cairns (Finn. *hiidenkiuas*), are the main marker of

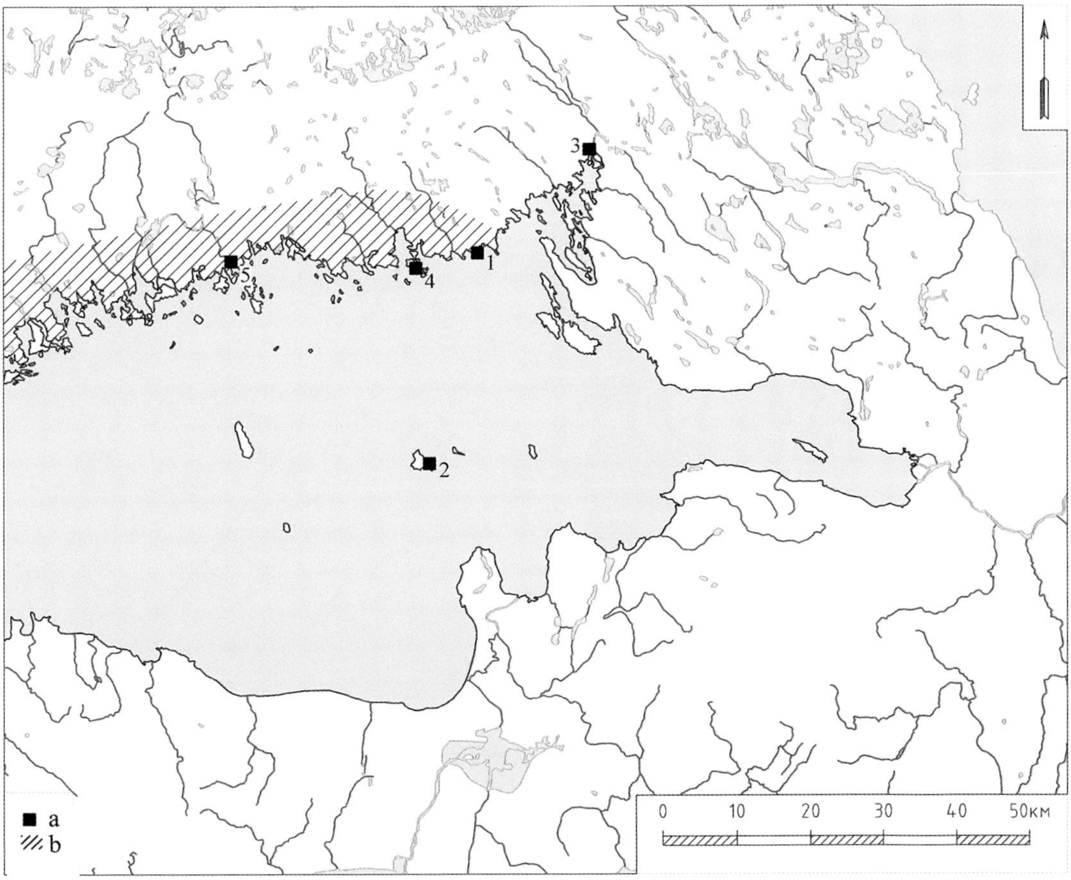

Figure 1. Antiquities of the Bronze Age on the northern coast of the Gulf of Finland. a — antiquities of the Bronze Age, b — area of the distribution of the North-European culture of the Bronze Age in the eastern part of the Gulf of Finland. 1 — cairns of Bolshoy Bor, 2 — cairn of Suisaari, Moshchny Island (Lavansaari), 3 — a single find, Vyborg (Tiikanurmi), 4 — cairn of Salovaara, 5 — cairn on Bolshoy Pogranichny island (Paatio). Drawing A. Gorodilov.

this archaeological culture. Currently, there are about 10000 known cairns in Finland, with a major part of them dating back to the Bronze Age (Salo 1992: 6; Tuovinen 2002: 66), with more than 3000 of them having been recorded (Aspund 2008: 72; Tuovinen & Vuorinen 1992).

In Finnish literature, the eastern boundary of the area of the culture of Bronze Age stone cairns is generally placed at the eastern coast of the Gulf of Finland, i.e. Russia's Leningrad oblast (e.g. Lavento 2003; Meinander 1954: 116–117) (Fig. 1, b). This opinion is based on the information gathered by Aarne M. Tallgren regarding five Bronze Age stone cairns in the vicinity of Bolshoy Bor (the village of Jokikylä in the locality of Kainiemi, on Laivakari rock) (Tallgren 1907: 72). An examination of these cairns was conducted by R. Rosén in 1917–1920.

In addition, the stone cairns on the Outer Islands of the Gulf of Finland contain information that possibly dates them to the Bronze Age. These data are summarised by T. Miettinen (Miettinen 1996). On Hogland (Suursaari) island, two cairns about 5 m in diameter were recorded on the heights of Lounatkorkia and Kumpelkallio. There is infor-

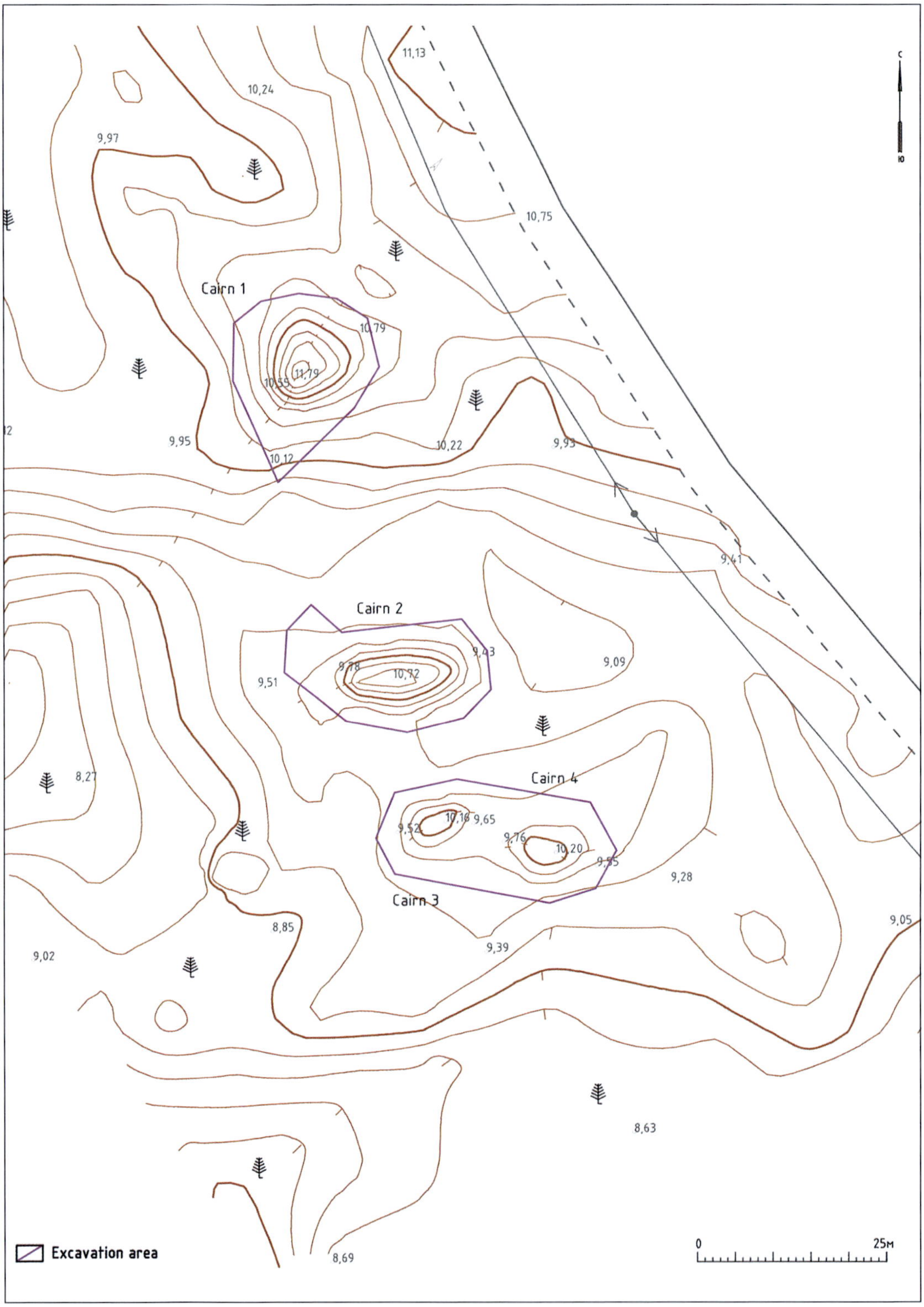

Figure 2. Topographic plan of the cairn group near the village of Bolshoy Bor. Drawing A. Gorodilov.

mation that a local teacher named E. Elenius carried out excavations of the Kumpelkallio cairn, where he found pieces of charcoal and burnt bones.

According to Torsten Edgren (Edgren 1993), a local teacher named Johannes Suomalainen informed the Archaeological Commission of Finland in 1926 that he had uncovered a stone cairn on the Suisaari promontory of Moshchny Island (Lavansaari). His measurements showed that the cairn was 10 m in diameter and 1,5 m high; nearby was a stone circle with a diameter of 7,5 m (Edgren 1993; Uino 1997: 269–270).

Along with the stone cairns, an object also dated to the Bronze Age was discovered, consisting of a single find of a spectacle fibula from the surroundings of Vyborg (found in Tiikanurmi, fig. 1, 3) (Hackman 1897: 43, Fig. 31; Meinander 1954: 116–117). This brooch belongs to type XXIV E2, according to Evert Baudou, and is encountered in Northern Europe from periods IV and V (i.e. ca 1100–700 BC) (Baudou 1960: 75–76, Taf. XV). Andreas Oldeberg, in his study on spectacle fibulae, dates the Vyborg find to period V (about 900–700 BC) (Oldeberg 1933). The nearest finds of spectacle fibulae come from southwestern Finland and a hoard from western Estonia (Saaremaa) (Sperling 2014: Abb. 4, 5).

3 Excavations of the cairns near the village of Bolshoy Bor in 2017

The group of cairns of 'Bolshoy Bor 1' is situated in the Vyborg region of Leningrad oblast within the territory of the rural settlement of Seleznevo (Ykspää), 2 km southwest from the settlement of Bolshoy Bor (Jokikylä, Säkkijärvi district), in a forest 30 m westward from a country road leading to Cape Konek (Koiniemi) and 1,5 km northward from the extremity of the cape (Fig. 1, 1). The cairns are located on the rock outcrop ("*ram's forehead*") overlooking the sea, ranged in the submeridional direction and surrounded by boggy lowlands on three sides. The absolute altitude of the rock surface here ranges between 9,5 and 11 m above sea level. The earliest information on the stone cairns on Cape Konek dates back to the beginning of the 20th century; this group was re-discovered by Aleksandr M. Zhul'nikov in 2016.

The group under consideration consists of four cairns arranged in a chain in the submeridional direction and distanced 4–15 m from each other (Fig. 2). Each cairn was tied to a local elevation of the rock surface 0,2–0,7 m high. This element creates the impression of the cairns being higher than they actually are. E.g., the structure of cairn no. 2 has two courses in height that is about 0,5 m, but visually the seeming height of the cairn is at least 1 m.

In 2017, a regional Leningrad Oblast expedition of the Institute for the History of Material Culture (IIMK) of the Russian Academy of Sciences (RAS) carried out explorations at the site. Four excavations were started, covering a total area of about 1050 sq. m. Cairns no. 1, 3 and 4 were investigated completely, while cairn no. 2 was studied partially.

Cairn no. 1 is located in the northern area of the group, on a rocky elevation at height marks of ca 11 m above sea level; it was partially covered with moss and overgrown with a few trees. After the moss layer and trees had been removed, the general structure of the object was revealed. The cairn under study was an amorphous stone cairn measuring 16 × 17 m and up to 1,3 m in height; in its central part, a depression up to 0,3 m deep was found. From the south, an additional structure measuring 4 × 7 m adjoined the cairn. The total area of the cairn was 200 m^2. A wall of a round plan constructed of large granite boulders served as the base of the cairn. The boulders rested upon the layer of the buried soil (Fig. 3, 4). The external diameter of the wall was 10,5 m; the stones were set tightly to each other and measured about 0,7 × 0,5 × 0,5 m. In the northeastern section of the cairn, a second course of the stone wall was recorded; these lay upon the first course and were of smaller dimensions. It is probable that the circle was origi-

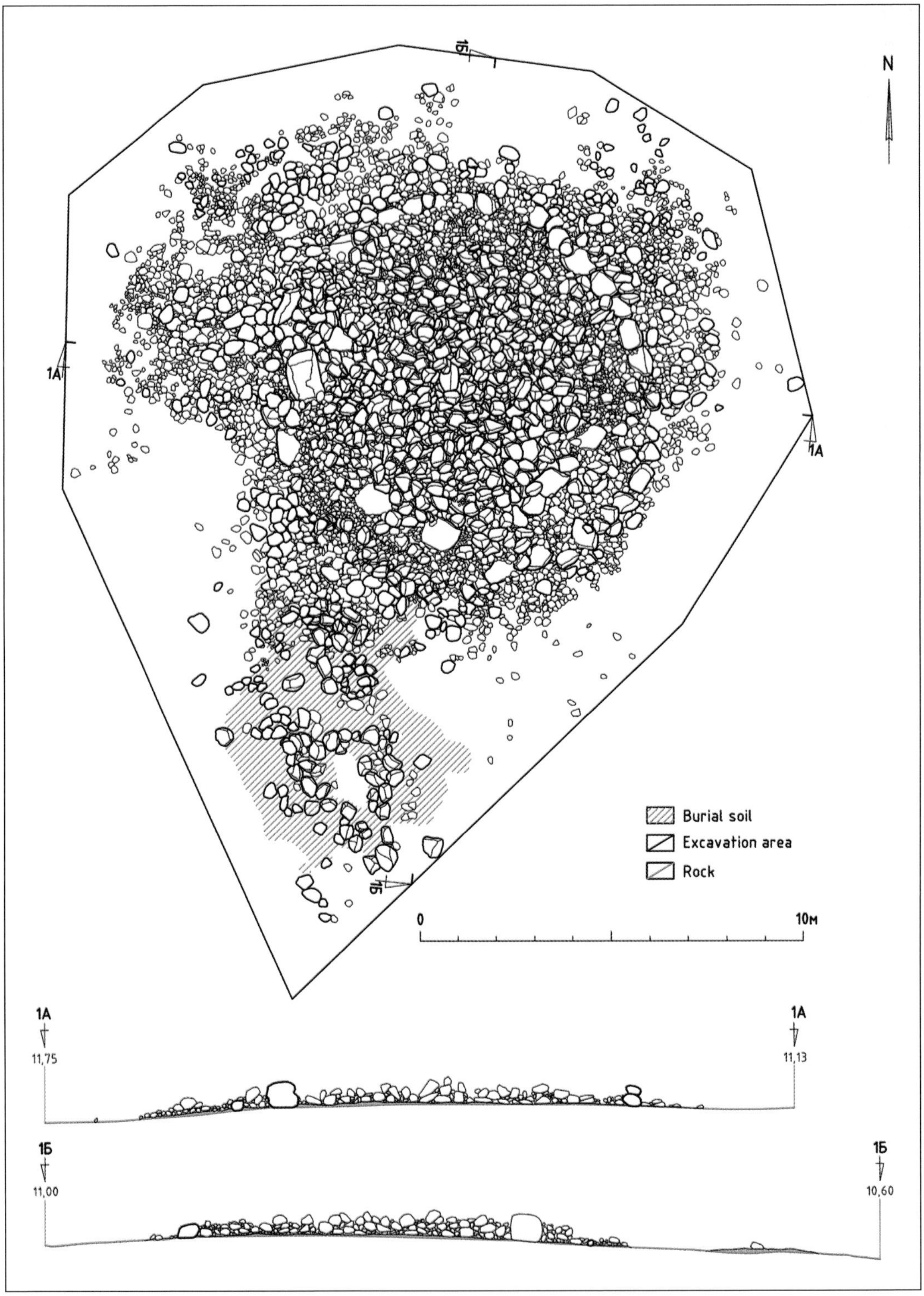

Figure 3. Plan of cairn 1 after removing the topsoil layer. Drawing A. Ivanova, A. Gorodilov.

Figure 4. Cairn 1 after partial dismantling of the cairn. Photo: A. Gorodilov 2017.

nally constructed of several courses of stones which eventually were scattered, forming a chaotic collapse outside the central circle. In the internal area, in the southern half of the cairn, there was a concentric semi-circle constituted by massive boulders (Fig. 5). The lower horizon of the inner space of the cairn is constituted by tightly laid medium-sized boulders. Around the inner side of the wall is a 'belt' about one metre wide consisting of smaller boulders and pebbles lying over the lower horizon. Under the lower horizon, an intercalation of buried soil with a thickness of 5–7 cm was found covering the rock foundation.

Excavation of the buried soil layer revealed three concentrations of burnt bones, all of them in the upper part of the layer. Concentration 1 was located in the eastern part of the cairn, to the north of the inner half-ring. The dimensions of the concentration are 0,9 x 0,6 m, while the area is 0,3 m^2. The bones in the concentration were poorly burned. In total, about 500 fragments of burnt bones with a size exceeding 0,5 x 0,5 cm were found in the concentration. The maximum size of the fragments does not exceed 2 x 1 cm, with the average size being 1 x 0,5 cm.

Concentration 3 was revealed at a distance of 0,5 m from concentration 1, also on top of a layer of buried soil. Bone concentration was overlapped with massive stones of the inner half-ring. The dimensions of the stones are 1,2 x 1 x 0,3 m and 1 x 0,8 x 0,5 m. The bones in the concentration were located in two compact groups of 0,3 x 0,3 m, located at a distance of 0,5 m from each other; single burnt bones were scattered in this space. In total, 250 bone fragments with dimensions of more than 0,5 x 0,5 cm were found. The maximum size of the bones does not exceed 2 x 1 cm.

Concentration 3 was probably the earliest one, made at the time of the construction of the cairn. Concentration 1 probably be-

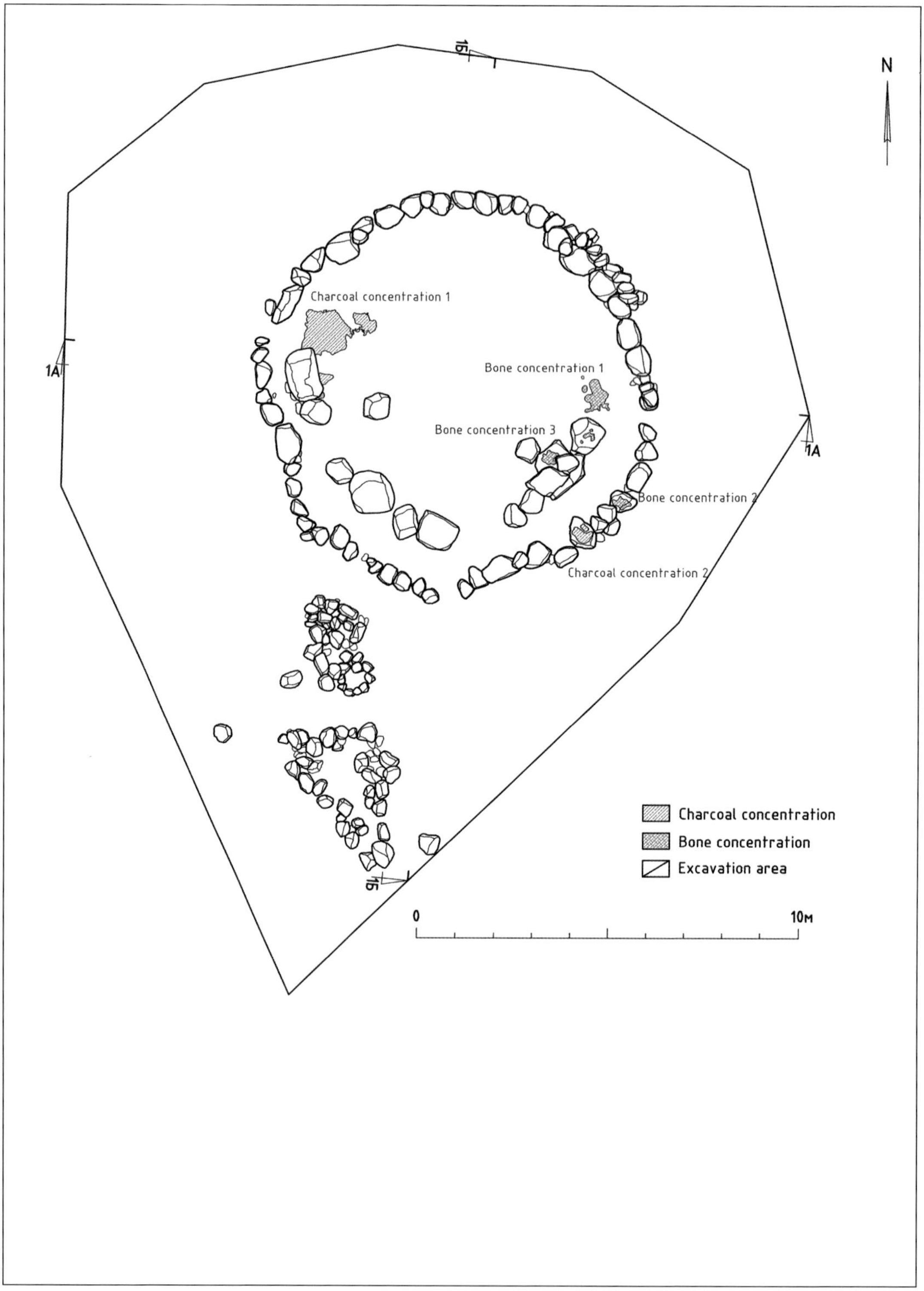

Figure 5. Plan of cairn 1 after the dismantling of the cairn. Drawing A. Ivanova, A. Gorodilov.

longs to the same period, since no traces of the reconstruction of the stone fill above it was noted.

Bone concentration 2 was found under one of the stones of the ring structure in the southeastern part of the cairn. The concentration was also corresponding to the upper part of the buried soil layer. The size of the concentration is 0,5 x 0,3 m. In total, it consisted of 30 fragments of burnt bones with dimensions not exceeding 1 x 1 cm. A massive boulder with dimensions of 0,7 x 0,5 x 0,5 m was located above the concentration.

The bones in all three concentrations are highly fragmented and relate mainly to tubular bones. In the course of the analysis carried out at the Museum for Anthropology and Ethnography (RAS), it turned out to be impossible to determine their species.

Apart from the bone concentrations, a concentration of coals measuring 0,5 x 0,4 m was revealed in the southeastern part of the stone ring lying on a layer of buried soil under a boulder measuring 0,9 x 0,6 x 0,5 m. Another concentration of coals measuring 1,1 x 1,2 m was revealed in the western part of the cairn to the north of the inner half-ring, close to a huge stone 1,3 x 0,8 x 0,8 m in size. The latter was partially crumbled, probably due to fire impact.

^{14}C dates of the 13th and 19th centuries were obtained from the bones of concentration 1 and the buried soil layer samples of cairn 1[1]. The late dating is probably the result from a large amount of modern organics, including tree roots.

Cairn no. 2 is located in the centre of the group on a rocky elevation up to 0,7 m in height at the height marks of about 10 m above sea level; it was partially covered with moss and also overgrown with a few trees. The cairn was an amorphous stone cairn measuring 12 x 14 m and up to 1 m high. The total area of the cairn was 170 m^2 (Fig. 6, 7). An excavation of the cairn was not carried out on account of the planned conservation of the site as an object of archaeological heritage. A stone annexe on the western side of the cairn was excavated. The annexe had dimensions of 3 × 7 m and was constructed of 1–2 courses of medium- and small-sized boulders. In the foundation of the annexe, two circular stone structures about 2 m in diameter adjoining tightly with each other were cleared out. In the eastern circular structure, a concentration of burnt bones covering an area of 0,3 x 0,3 m was uncovered. In total, about 25 fragments of burnt bones with dimensions exceeding 0,5 x 0,5 cm were found in the concentration. The maximum size of the bones did not exceed 1 x 1 cm. No fragments from the concentration were identified.

Cairns no. 3 and 4 are located in the southern section of the group on small rocky elevations up to 0,2 m high, at height marks of about 10 m above sea level. The distance between the cairns was 4 m; they were investigated over a single excavated area.

Cairn no. 3 was represented by a nearly round structure measuring 9 x 11 m and with a height of about 0,5 m; it was partly covered with moss. The total area of the cairn was 75 m^2. After dismantling the upper horizon of chaotically lying stones (Fig. 8, 9), the internal structure of the cairn was revealed. The lower horizon was constructed of stones of nearly equal dimensions (on average 0,4 x 0,3 x 0,3 m) laid tightly to each other. On the northern, eastern and western sides around the outer edge of the cairn, these stones formed something like a wall bordering the internal space. However, no distinct structure was discernible as it was in the first cairn. In the southwestern section of the inner space, a cupstone was found; in its centre was a round depression 3 cm in diameter and 4 cm deep, possibly of natural origin. In the central part of the cairn was a vertically set stone block with a height of 0,5 m. To the east of the centre of the cairn, in the upper part of the layer of buried soil beneath the stones, was a concentration of burnt bones measuring 0,5 x 0,6 m. In total, about 110 fragments of burnt bones exceeding 0,5 x 0,5 cm in size were found in the concentration. The maximum size of the bones did not exceed 2 x 1

Figure 6. Cairn 2 after removing the topsoil layer. Photo: A. Gorodilov 2017.

cm. No fragments of bones were identified (Fig. 10).

Cairn no. 4. From the beginning of the excavations, the eastern part of this cairn was disturbed and the western part was partially covered with moss. In terms of its shape, this cairn was similar to cairn no. 3; it was a nearly round structure measuring 12 x 9 m and about 0,4 m high. The total area of the cairn was 75 m^2. Before the beginning of the excavation, the nature of the internal structure of the cairn was unclear (Fig. 8, 9). After dismantling the stones chaotically lying in the peripheral section of the cairn, a structure about 5 m in diameter was revealed. The lower course of the masonry was laid of stones of a similar size (about 0,4 x 0,3 x 0,3 m) laid tightly to each other. Beneath the stones, in a layer of the buried soil to the north from the centre of the cairn, a concentration of burnt bones measuring 1,5 x 0,7 m was extended in the latitudinal direction. The bones were located in the upper part of the layer of buried soil. In total, about 200 fragments of burnt bones exceeding 0,5 x 0,5 cm in size were found in the concentration. The maximum size of the bones did not exceed 2 x 1 cm. No fragments of bones were identified (Fig. 10). In the central area of the concentration, a bronze knife was uncovered among the burnt bones (Fig. 10, b). The knife is rather small, with a length of 6.1 cm and the blade being 1 cm wide, 2 mm thick, and triangular in the cross-section; its edges were worked through the technique of cold hammering. The knife was manufactured from a copper-based alloy with an admixture of 6–8% tin, 2–3% arsenic, 1–2% antimony, 0,5–1% nickel, less than 0,6% silver and less than 0,5% zinc.[2]

4 The chronology and cultural context

The excavated cairns constitute a single complex. They are each tied to identical elements of the landscape, these being small elevations on the rock's surface, and are located within the limits of a single rock. They share a series of common construction features. The outer outline of the cairns is bordered with a stone wall; the latter is best preserved in cairn no. 1 (masonry of two courses); the diameter of these circular structures varies from 5 m (cairn no. 4) to 10,5 m (cairn no. 1). The amorphous outlines of the cairns observable prior to the excavations had probably been formed due to the collapse of the stones of the external walls, whereas originally the cairns had vertical walls and were of a round plan. The lower horizon of the internal space of the cairns was constituted by tightly laid boulders of similar size. Different stone structures were recorded in the foundations of the cairns: a semi-circle in cairn no. 1 and an upright stone in the centre of cairn no. 3. Of note is also the discovery of a cupstone in the construction of the lower horizon of cairn no. 3. One to three concentrations of burnt bones and charcoal were found in the layer of the buried soil beneath the lower horizon of stones in each structure. A bronze knife was uncovered in one of the bone concentrations. Two of the cairns were adjoined by stone masonries. Analysis of the burnt skeletal remains has not allowed us to identify their species.

It is now believed that the construction of stone cairns in Finland could have started as early as the end of the Stone Age. This theory is supported by ^{14}C dates obtained for cairns from the Finnish interior (Taavitsainen 2003a; 2003b). However, the main part of the cairns belongs to the Bronze and Iron Ages. Until 1988, a total of 250 cairns were excavated in Finland. Of these, 114 were dated: 57% to the Bronze Age, and the rest to the Iron Age (Asplund 2008:73).

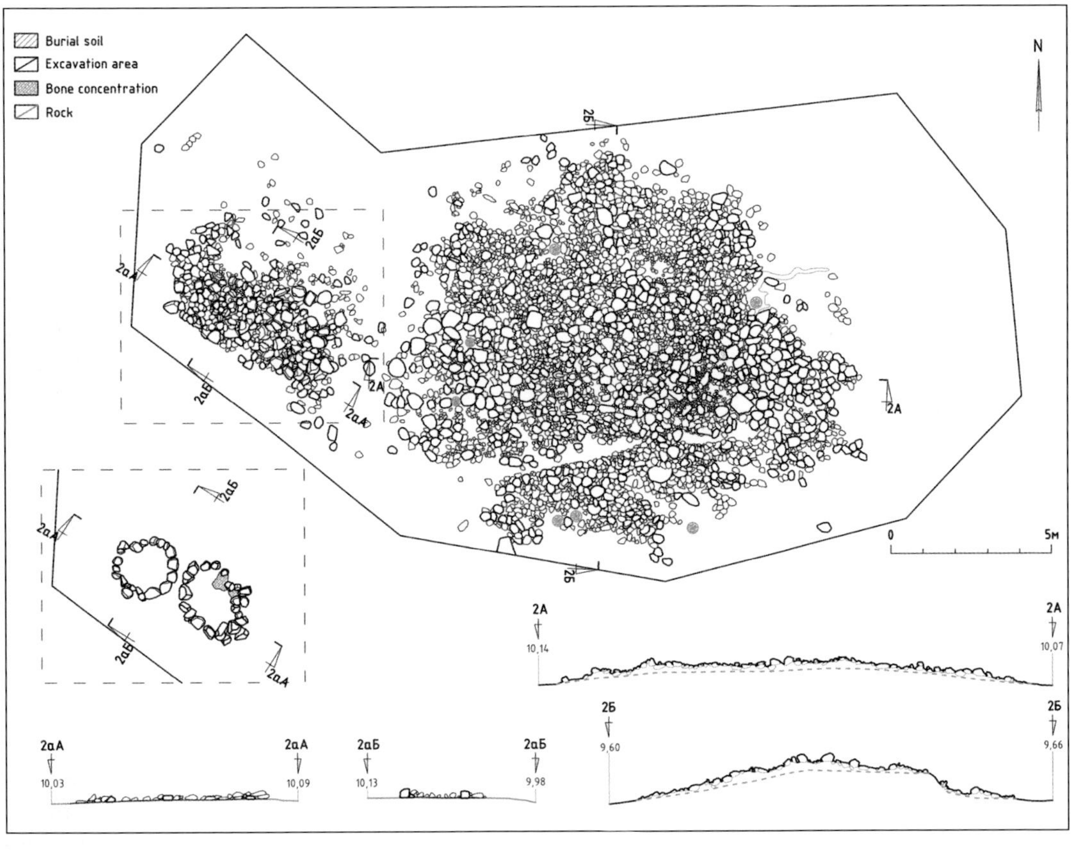

Figure 7. Plan of cairn 2 after removing the topsoil layer. Drawing A. Ivanova, A. Gorodilov.

Figure 8. Cairns 3 and 4 after removing the topsoil layer. Photo: A. Gorodilov 2017.

The study showed that a significant portion of the cairns belonged to the Iron Age. T. Tuovinen identified two groups: P (Bronze Age cairns) and R (Iron Age cairns) (Tuovinen 2002: 190-195). A detailed analysis of the construction time of the cairns was made on the basis of morphological and spatial criteria by T. Tuovinen while studying the cairns of the Åboland region in southwestern Finland. He found that a significant part of the cairns belonged to the Iron Age.

An important difference between the identified groups lies in the average area of the stone covering of the cairn. The average area of the cairns of group P reaches 162 m^2, while the average area for group R was 28 m^2. Also, Tuovinen identified microtopographic characteristics of the location of the cairns as defining features. The cairns of group P are usually located low in relation to the highest top, but still on a hill, whereas the cairns of group R are more often high in relation to the highest top but located on the plane (Tuovinen 2002: 193).

Among the sites excavated in various regions of the coastal part of Finland, the closest analogies for the cairns near Bolshoi Bor are cairns no. 86 and 98 at the cemetery of Rieskaronmäki and the burial ground of Uotinperä 2 in Nakkila in the Satakunta region investigated by Unto Salo (Salo 1981: 64–95, kuva 20, 39, 40). In one of the burials, a bronze knife similar to the find from cairn no. 4 at Bolshoy Bor was found (Salo 1981, kuva 117). The knife is 6,27 cm long, with the blade being 1 cm wide.

A similar cairn was explored by J. Pukkila in 2004 at Alastaro Isovarenmäki in southwestern Finland (Pukkila 2004; Pukkila & Pellinen 2019). In a burial cairn measuring 8 x 10 m, Pukkila found a concentration of burnt bones measuring 1,3 x 1 m with a fragment of a bronze plate and a fragment of a bronze knife ca. 2 cm long, 1 cm wide and 2 mm thick (Pukkila & Pellinen 2019, kuva 10, 11). This knife fragment is the closest analogy to that found at the excavation of cairn no. 4 near Bolshoy

Bor. They are similar in both metric characteristics (width, thickness, section) and shape. Knives of this type are known from the excavations of the sites of Denmark and southern Sweden in the II-IV periods (Larsson 1986: 44-45, Pukkila & Pellinen 2019:111); at least five such knives are known in Finland[3]. For the bones from Alastaro Isovarenmäki, the ^{14}C date of 2835 ± 45 BP (Hela-896) was obtained, which allowed the construction of the cairn to be dated to between the late 2nd and early 1st millennium BC. Based on the date and the knife find, J. Pukkila identified the cairn to be from the IV period of the Bronze Age (Pukkila & Pellinen 2019: 108-110).

The closest Bronze Age burial cairns to Bolshoy Bor are located in the region of Kymenlaakso in southeastern Finland. Currently, about 50 cairns are known to be there (Miettinen 2012:55). The closest excavated cairn reliably dated to the Bronze Age is that of Salovaara, investigated by T. Miettinen in 1991–1992 near Kotka (Miettinen 1997). This cairn is located about 70 km to the west from the cairns of Bolshoy Bor (fig. 1, 4). Here, a cairn of a similar form, with a diameter of ca. 11 m and with an annexe was examined. It is situated on an altitude of 10,5 m (Baltic system), similarly to the Bolshoy Bor cairns. During the excavation, a double spike button of the North-European type dated to period IV, according to Baudou (ca 1100–900 BC), was found in a concentration of burnt bones (Baudou 1960: 87–88, Taf. XVIII: XXVIA1). The ^{14}C date for the bones, 700 ± 30 BP (Su-2805), refers to the 13th-14th centuries AD, likely because of the modern vegetation in the sample (Miettinen 2012:60).

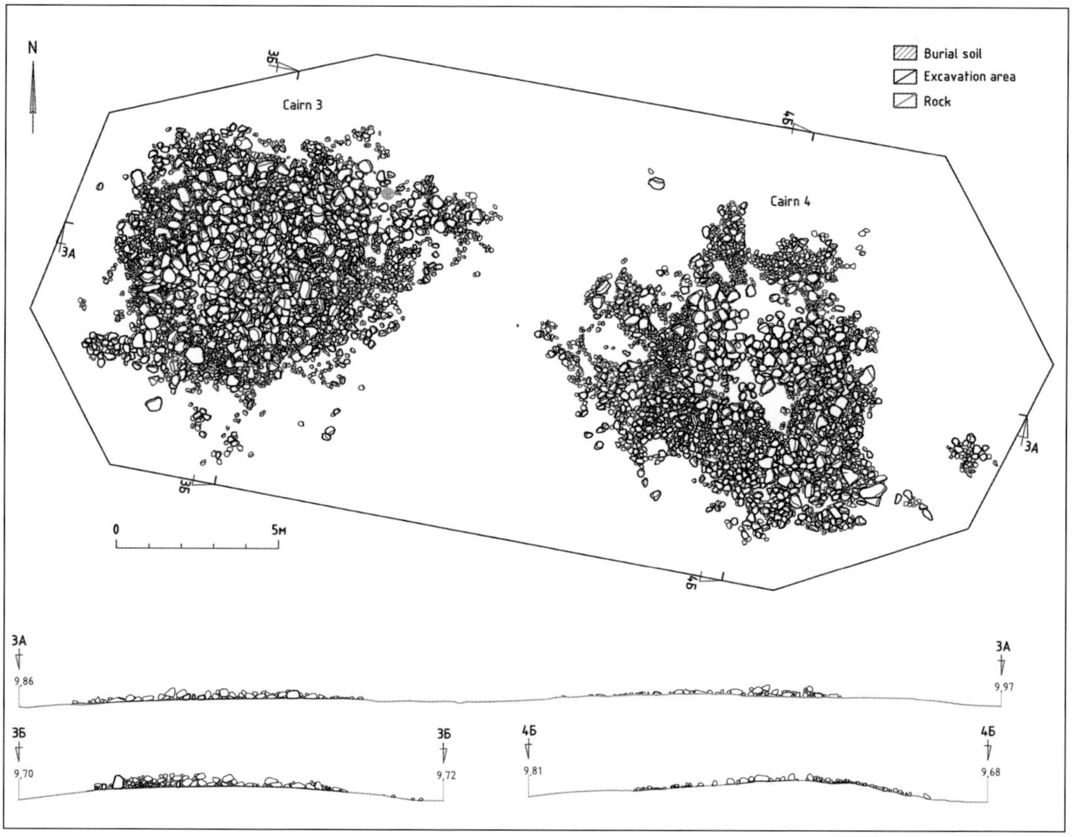

Figure 9. Plan of cairns 3 and 4 after removing the topsoil layer. Drawing A. Ivanova, A. Gorodilov.

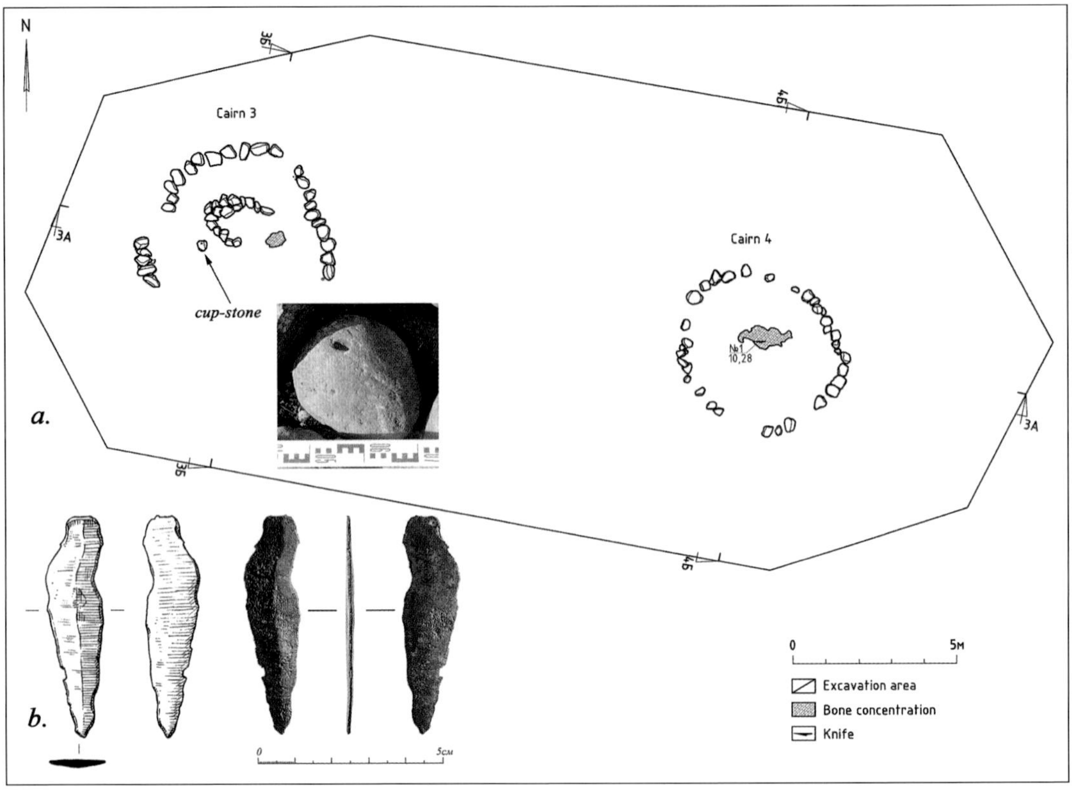

Figure 10. a. Plan of cairns 3 and 4 after the dismantling of the cairn. b. Bronze knife from cairn 4. Drawing A. Ivanova, A. Gorodilov.

In 1939, Ella Kivikoski carried out excavations of the stone cairn on Bolshoy Pogranichny island (formerly Paatio island) in Virolahti Bay (Kivikoski 1940). This cairn is situated 20 km to the west from the cairns of Bolshoy Bor (Fig. 1, 5). The cairn is located on a rock, at a height of 6,2 m above sea level; it is 6–6,5 m in diameter and about 0,75 m high; in its centre, the cairn was disturbed by a pit. During the excavations, a circle structure laid of larger boulders 3,5 m in diameter was revealed in the foundation of the stone cairn. Under the layer of stones, burnt bones (36 g) and a few fragments of flint were collected from the buried soil. E. Kivikoski estimated that, through its form and structure, this cairn must be dated to the Bronze Age, but she also did not exclude the possibility of a younger date based on the altitude chronology of the shore terraces.

5 Conclusions

The cairns near the village of Bolshoy Bor can be dated to the Bronze Age for a number of reasons. The cairns are located at an altitude of 9,5-11 m (Baltic system) and on a rocky hill, but not at the dominant height of the terrain, which is typical for the cairns of the Bronze Age. All cairns are located on local rocky hills. This makes them appear larger than they actually are, which is also a feature of the Bronze Age burial cairns (Holmblad 2010:121). The area of each cairn near Bolshoy Bor ranges from 75 to 200 square metres, which also allows us to attribute them to the Bronze Age. The internal structure of the burial mounds, namely the presence of ring stone constructions and annexes (cairns no. 1 and 2), have analogies among the Bronze Age burial cairns located in neighbouring territories, including the burial grounds of Rieskaronmäki, Watinpera 2, Alastaro Isovarenmäki, Salovaara, etc. Unfortunately, the ^{14}C dates of the bones from the concentrations did not provide accurate dating of the construction, probably due to the presence of modern organics, including tree roots, in the samples. However, the bronze knife found in cairn no. 4 is closely similar to that found in Alastaro Isovarenmäki and can also be attributed to the Bronze Age IV period. The question of the duration of the life of the site cannot be finally determined. It is likely that cairns no. 3 and no. 4 did not function for long, judging by the absence of a complex structure of the cairns and the single concentrations of bones. Cairn no. 1 could have functioned for a longer time. Both the presence of an inner half-ring inside the cairn, with the initial bone concentration under its stones, and the presence of bone concentration under the stones of the outer ring suggest the possibility of a burial conducted at a later time. Similar cases are known in Finland (Pukkila & Pellinen 2019:110-111). Taking into account these facts, the period of function of the cairn group near Bolshoy Bor can be limited to IV-V period (the late 2nd – early 1st millennium BC).

The coastal region of the Gulf of Finland west of Vyborg (within Russian territory) has remained a rather poorly studied area until recently. The continuation of archaeological investigations will allow the researchers to define more precisely the chronological and territorial boundaries of the tradition of the construction of stone cairns, or *hiidenkiuas*, and to examine in more detail the dynamics of the cultural interactions in the region of the Gulf of Finland during the Bronze Age.

References

Archival sources

Kivikoski, E. 1940. Kertomus röykkiön kaivauksesta Virolahden Paation Västvikenissä heinäkuun 5 p:nä. 1939 (https://www. kyppi. fi/palveluikkuna/raportti/read/asp/hae_liite. aspx?id=113617&ttyyppi=pdf&kansio_id=935. Read 20.03.2020)

Pukkila, J. 2004 Alastaro Isovarenmäki pronssikautisen hautaröykkiön arkeologinen tutkimuskaivaus. 2004. (https://www. kyppi. fi/palveluikkuna/raportti/read/asp/hae_liite. aspx?id=101753&ttyyppi=pdf&kansio_id=430. Read 20.03.2020)/

Literature

Asplund, H. 2008. *Kymittæ. Sites, centrality and long-term settlement change in the Kemiönsaari region in SW Finland*. Annales Universitatis Turkuensis B 312.

Baudou, E. 1960. *Die regionale und chronologische Einteilung der jüngeren Bronzezeit im Nordischen Kreis*. Acta Universitatis Stockholmiensis. Studies in North-European Archaeology. I. Stockholm: Alkvist & Miksell.

Edgren, T. 1993. Lavansaaren Suursuonmäen röykkiöhaudat. *Suomen Museo* 1992: 5–20.

Hackman, A. 1897. *Die Bronzezeit Finnlands*. Suomen Muinaismuistoyhdistyksen Aikakauskirja XVII.

Holmblad, P. 2010. *Coastal Communities on the Move. House and Polity Interaction in Southern Ostrobothnia 1500 BC – AD 1*. Archaeology and Environment 26. Umeå

Larsson, T. B. 1986. *The Bronze Age Metalwork in Southern Sweden. Aspects of social and spatial organization 1800-1500 B.C.* Archaeology and Environment 6. University of Umea, Dept. of Archaeology, Umeå

Lavento, M. 2003. Viipurin läänin pronssikausi ja varhaismetallikausi. *Karjalan synty*: 245-290. Juvaskyla: Karjalan kirjapaino.

Meinander, C. F. 1954. *Die Bronzezeit in Finnland*. Suomen Muinaismuistoyhdistyksen Aikakauskirja.

Miettinen, T. 1996. Suomenlahden ulkosaarten esihistoria. In *Suomenlahden ulkosaaret – Lavansaari, Seiskari, Suursaari, Tytärsaari*: 52-67. Helsinki.

Miettinen 1997 = Миеттинен, Т. 1997. *Изменение картины железного века на северном побережье восточной части Финского залива. Славяне и финно-угры. Археология, история, культура*. Доклады российско-финляндского симпозиума по вопросам археологии: 62-70. Санкт-Петербург.

Miettinen, T. 2012. Kymenlaakson esihistorian kehitys. In Y. Kaukiainen (ed.) *Kymenlaakson historia 1, Jokilaakso ja rajamaa esihistoriasta 1810-luvulle*. Helsinki: Suomalaisen Kirjallisuuden Seura.

Oldeberg, A. 1933. *Det nordiska bronsålderspännets historia*. Stockholm.

Pukkila, J. & Pellinen, H.-M. 2019. Alastaron Isovaremäen muinaisjäännös – Lapinraunio vai hiidenkiuas?

In Harjula, J., Immonen, V. & Ruohonen, J. (eds.) *Puukenkien kopinaa. Henrik Asplundin juhlakirja*. Karhunhammas 19: 101–122.

Salo, U. 1981. Satakunnan historia I, 2. *Satakunnan pronssikausi*. Rauma: Satakunnan Maakuntaliitto r.y.

Salo, U. 1992. Projekti Lounais-Suomen rannikon varhaismetallikautiset hautarauniot. *Karhunhammas* 14: 5-8.

Sperling, U. 2014 *Aspekte des wandels in der Bronzezeit im Ostbaltikum die siedlungen der Asva-Gruppe in Estland*. Tallinn: Eesti Teaduste Akadeemia Kirjastus.

Taavitsainen, J.-P. 2003a. Lapinraunioiden kronologisfunktionaalisten kysymysten hahmottelua – uusia AMS-ajoituksia Keski-Suomen lapinraunioiden palaneesta luusta. *Muinaistutkija* 1/2003: 2-23.

Taavitsainen, J.-P. 2003b. Lapp Cairns as a Source on Metal Period Settlement in the Inland Regions of Finland. *Acta Borealia* 1/2003: 21-47.

Tallgren, A. M. 1907 Kivikautinen taideteos Säkkijärveltä. *Suomen Museo* 1907: 67–72.

Tuovinen, T. 2002. *The Burial Cairns and the Landscape in the Archipelago of Åboland, SW Finland, in the Bronze Age and the Iron Age*. Acta Universitatis Ouluensis B Humaniora 46.

Tuovinen, T. & Vuorinen, J.-M. 1992. Luettelo Suomen rannikon hautaraunioista. *Karhunhammas* 14: 9-122.

Uino, P. 1997. *Ancient Karelia. Archaeological studies – Muinais-Karjala. Arkeologisia tutkimuksia*. Suomen Muinaismuistoyhdistyksen Aikakauskirja 104.

Notes

1 770+-25 BP (1220-1280 AD (95.4%)) (SPb-2526) and 80+-25 BP (1692-1729 and 1811-1920 AD (95.4%)) (SPb-2519).

2 The composition of the metal was identified by Sergey V. Khavrin (The State Hermitage Museum).

3 I would like to thank Jarkko Saipio for his help in attributing and searching for analogies to the knife from the Bolshoy Bor site.

The Long-Period Dwelling Site of Multavieru at Lake Höytiäinen in North Karelia – Approaching Questions in the Municipalities of Polvijärvi, Kontiolahti, Joensuu and Liperi

Mika Lavento

Mika Lavento, University of Helsinki, Department of Cultures, Archaeology, P.O. Box 59 (Unioninkatu 38F), FI-00014 University of Helsinki, Finland, e-mail: mika.lavento@helsinki.fi

Abstract

The starting point of the paper is the dwelling site of Multavieru, located near the Kiskonjoki River, on the NW side of Lake Höytiäinen in North Karelia. The remains of a dwelling-site area were found in a zone of ca 350 m long and not larger than 80 m wide near the Kiskonjoki River. Multavieru was discovered in 1969 and it has been investigated several times; in addition, it was studied twice by way of teaching excavation by the department of archaeology, the University of Helsinki. The habitation at the site covers the whole Neolithic and Early Metal Age times. A considerable part of the finds implied remains of Medieval Period and iron smelting dating to the 17th and 18th centuries AD. The article illuminates the prehistory and changes in habitation near Lake Höytiäinen from the late Mesolithic to the 20th century AD. The second question addresses the sites close to Multavieru, together with their finds, because this indicates the ongoing changes of culture during the lengthy period under study. Finally, the article addresses the history of the area around Lake Höytiäinen, because many features in the culture have changed considerably, although the area itself has not been considered as very significant, when trying to understand the prehistory of Finland.

1 Introduction

The dwelling site of Multavieru is located near the Kiskonjoki River, on the NW side of Lake Höytiäinen (87,3 m asl.) in the province of Northern Karelia. The site was discovered in 1969 and has been investigated several times; in addition, it has also been excavated two times by the University of Helsinki's Department of Archaeology. The habitation in Multavieru covers the whole Neolithic and Early Metal Age. In the area exist remains of Medieval Period and iron smelting carried out at the same place during the 17th and 18th centuries AD.

The first aim in this presentation will be to illuminate the prehistory and research the changes in habitation near Lake Höytiäinen from the late Mesolithic to the 20th century AD. The second question will concentrate on the sites close to Multavieru itself. The prehistory will be investigated by studying and comparing, in general, the material found at the sites. In practice, the comparison will be done in four different municipalities – Polvijärvi, Kontiolahti, Joensuu and Liperi, and the sites known there. Finally, the investigation will discuss the water history of Lake Höytiäinen, due to the considerable changes in location of dwelling sites on the shores of the lake.

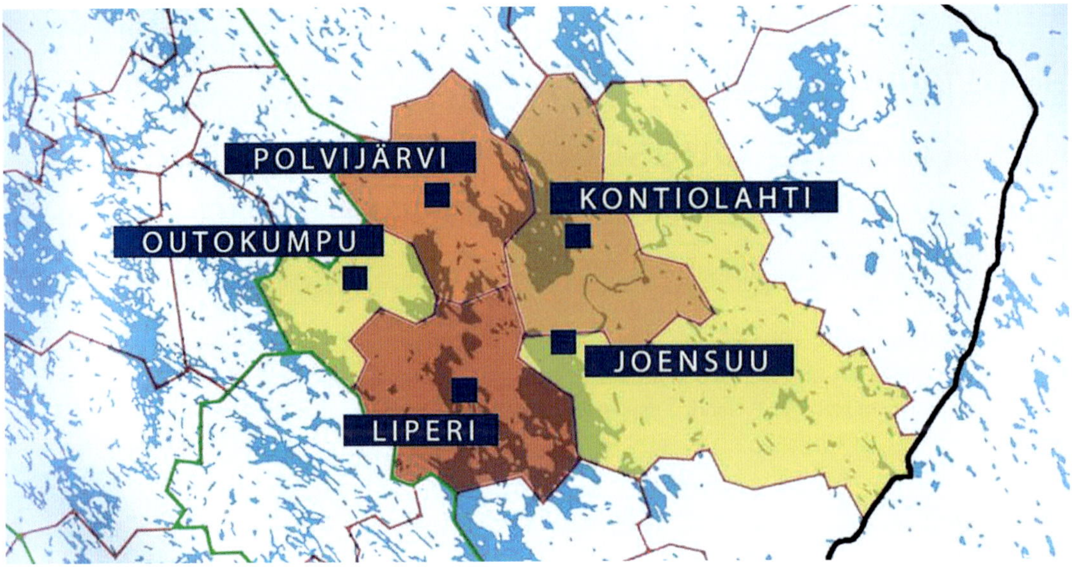

Figure 1. Polvijärvi, Kontiolahti, Joensuu and Liperi in northern Karelia. Outokumpu does not belong to the municipalities inspected in the article.

North Karelia and particularly of Lake Höytiäinen and the northern part of Heposelkä and northern Pyhäselkä in Lake Saimaa are located on the Svekofennian and Karelian Mica schist, Micaceous gneiss and Migmatite and, finally, on the eastern side of it, the Archaean bedrock (Ehlers et al. 1993). The Karelian mountains in eastern Finland developed ca 1920 million years ago. Eventually a copper mine was excavated at the site in Outokumpu (Koistinen 1981). Asbestos was in some places and in particular in Tuusniemi, on the NW side of Outokumpu. Its role in ceramics was central in the large area of Finland during the Middle Neolithic and the Iron Age (Lavento & Hornytzkyj 1996).

Questions

The article addresses three main questions, and the last one is additionally a sub-question. These will be discussed by starting with the results presented already in the lecture in Hämeenlinna in October 2019. Here, the questions will be approached in more detail. The context is slightly larger, too. The approaches are the following:

1. What were the changes of cultures and the habitation near Lake Höytiäinen in the period between the late Mesolithic and the 20th century AD?
2. How did the habitation develop in four municipalities in North Karelia?
3. Where and what are the sites around the site Multavieru in particularly?
4. What were the cultures and how did the material change in Polvijärvi municipality and its three neighbouring municipalities during the lengthy period under study?

The first question of the article addresses the site and living on the Multavieru site encompassing the four municipalities in North Karelia as well. The area is outlined by the municipalities of Polvijärvi, Kontiolahti, Joensuu and Liperi. In my article, the most important area for me is Polvijärvi. This area was emphasised because it had remained under minor investigation by the archaeologists in Finland – despite the fact that the area is large (Fig. 1) and it has been inhabited during a lengthy period of time from at least 7000 BC.

The second question the article addresses is the population in the area and its habitation. The purpose is to find out what the changes were that can be observed in the archaeological material – dwelling sites and the material – that took place in the research area in North Karelia during the time period. This was observed through the material, mostly in the Finnish Register of Ancient Monuments and Sites collected by the Finnish Heritage Agency.

The third question the article addresses is the archaeological excavation and the research of the multi-period dwelling site of Multavieru in the Polvijärvi municipality. The site was inhabited from the Stone Age to the Historical Period without longer breaks. Finally, the site was also used for other purposes during the 18th century, and these historical remains can still be seen there today. In addition, within the larger scope of these efforts, the intention was to carry out the research of soil, habitation and the environment of Multavieru and its neighbourhood (Kouki 1999).

The area has changed during the Holocene because the shore displacement meant that the lakes and rivers developed after the melting of the ice separating the smaller glacial lakes from the Baltic Sea when the ice retreated from Jaamankangas soon after 8 700 BC; the melting ice was in the northern part of Lake Höytiäinen (Pihlatie et al. 1997). The large lake became closed off from the Baltic Sea ca 9200 BC and, because the water of the lake was reduced in 1859 AD, it is at 87,3 m above the level of the Baltic Sea (Donner 1978). The size of the lake is today 282,64 km² and it has as many as 606 islands.

2 Lake Höytiäinen and other lakes

The research area of the article belongs to an environment, which has not researched by archaeological fieldwork in detail. In the area, the Archaeological Commission, Metsähallitus, and the North Karelian Museum briefly studied the municipalities. They all had reasons for fieldwork and in many cases the starting point of the work was land-use planning and, thereafter, building.

Polvijärvi municipality is located between Lakes Viinijärvi in the western part of the area, Pielinen in the east and Orivesi and Pyhäselkä in the northern part of Lake Saimaa. Much interest will concentrate on the area around Lake Höytiäinen and the archaeological observations there (Fig. 1). This is because the University of Helsinki carried out the inventory in the lake municipality, together with the students in 1996 (Lavento 1997) and two teaching excavations during 1996-1997, led by Mika Lavento (1997; 1999a). Some research was carried out at some sites in Polvijärvi municipality (Koivikko & Ruuttu 1996; Mikkola et al. 1998), and the first excavations at the sites have been done already in the 1950s. Among the most important dwelling sites belongs Sola Suovaara (Björkman 1961; Meinander 1962), where very early copper finds was discovered (Taavitsainen 1982), dated as early as ca 3800 calBC (Nordqvist & Herva 2013: 405).

Lake Höytiäinen belongs to the large inland lakes of eastern Finland that developed somewhat synchronously with the northern parts of Lake Saimaa (Fig. 2). However, it does not belong to Lake Saimaa but developed into its own lake in the area of the contemporary municipalities of Polvijärvi, Kontiolahti, Juuka and Kaavi soon after 7500 calBC (Saarnisto 1968). The lakes were studied by Väinö Auer and Matti Sauramo, who separated margin formations in the 1910s (Sauramo & Auer 1928; Donner 1978: 227-229). Antti Rietrikki Helaakoski and Aaro Helaakoski investigated the lake in 1914. However, they were able to uncover organic stratigraphic which implied "dead plans". The areas were studied in Vierevänniemi, which was known by Russian anarchist Pjotr Aleksejevitsh Kropotkin, the researcher of the shore displacement. The research was carried out by A. F. Thoreld, Botanist R. Herlin and Geologist W. W. Wilkman (Pihlatie et al. 1997).

At present, the main depth of the lake is 12 m, but the deepest depth is 65 m and the

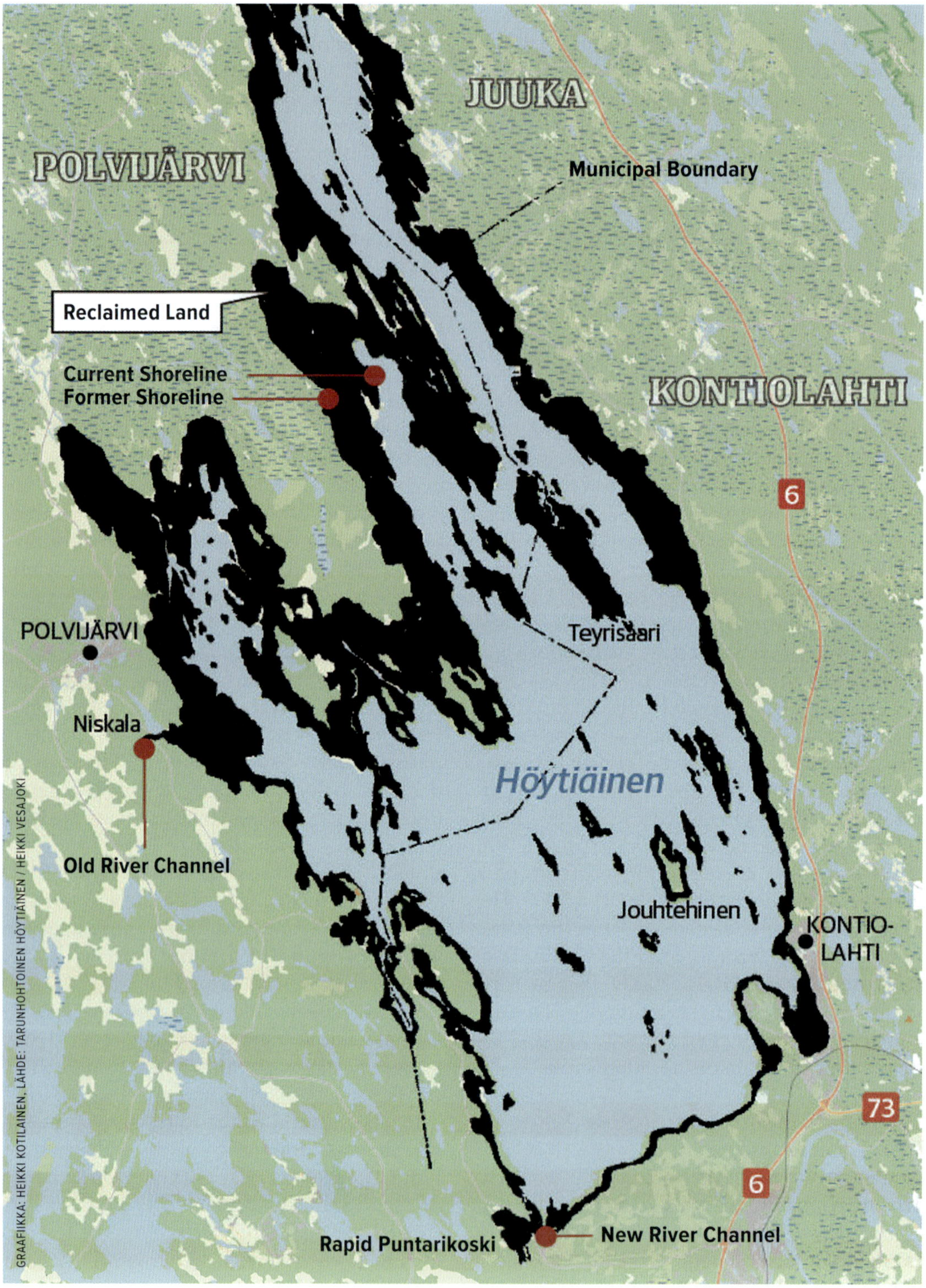

Figure 2. Developing of Lake Höytiäinen. Marked in black the naturally reclaimed land that formed as the result of the water lever drop down of Lake Höytiäinen after 1859.

open area of inland waters has a depth of 50 m. These deeper parts are located on the southern part of the lake and the bays are located on the NW part of the lake. This means that it is a relatively deep lake in Finland.

The most interesting observation about Lake Höytiänen is its history in the 17th century, when the surface of the water was lowered at the request of the local farmers.

In 1859, their group situated in Polvijärvi and Kontiolahti decided to dig a canal from Lake Höytiäinen to Lake Pyhäselkä in Saimaa. The original idea was to lower the water level some few metres. However, the operation did not work out as planned and the water level sank down 11,4 m. Today the highest water level is 75,9 m asl. in Lake Höytiäinen, which is the same as in Lake Pyhäselkä (Donner 1978: 228-229).

Another lake in the Polvijärvi municipality is Lake Viinijärvi, the depth of which is today 78,8 m above the level of the Baltic Sea. Its area is 134,91 km² and its deepest depth is 58,3 m, which means that it is a relatively deep lake, similar to Lake Höytiäinen. The western shores belong to the municipality of Outokumpu and the southern ones to Liperi; all other parts of the lake represent the municipality of Polvijärvi. The level of Lake Viinijärvi was lowered twice in recent historical times, first in 1830 and again in 1858, but altogether only 2 m (Anttila 1967). The water level of the lake is presently at 79,2 m asl.

One should observe that the dwelling sites around Lake Höytiäinen were situated on different elevations that proceeded the contemporary water level. The sites on the lowest elevation were found situated at 87 m asl. and the highest at ca 118 m asl.

3 Archaeological methods used in researching the dwelling sites in the municipalities

As already mentioned, the area in all four municipalities has mostly been archaeologically researched by surveys (Huurre 1985). Despite the surveys, however, the areas are not known in great detail. This is because the areas are large, and several of the targeted areas did not encompass any buildings, which could have served as starting points for detailed surveys or excavations at the sites. In addition, almost all municipalities have lost much of their population and have not had much interest in archaeological fieldwork.

The first general surveys were made in the area of Polvijärvi in 1965 (Sarvas 1965). The surveys were based on loose finds and checking the places where they were found. During the fieldwork, 13 new prehistoric monuments with sites were located. At the beginning of the surveys, one of the problems was that 1:20 000 maps earlier than the 1980s were missing. For instance, Pekka Sarvas and Pirjo Lahtiperä did not have the 1:20 000 maps in use, because this became possible in the area only after the 1980s. Kiihtelysvaara, Pyhäselkä and Tuupovaare areas experienced the same situation (Lahtiperä 1967a, b and c), in that they were separate municipalities before they were connected to the municipality of Joensuu in the early 2000s. In some cases, even the 1:10 000 maps were unavailable in Joensuu (Lahtiperä 1968a).

In North Karelia, a survey of the Early Metal Age and Historical sites was carried out during 1992-94 (Poutiainen et al. 1996) in all these municipalities discussed in the article. The problem of North Karelia in Finland was understood already by historian Heikki Kirkinen (1963: 11), who stated that the area has been insufficiently researched in relation to other areas of Finland. The survey added some information about the later periods, but the real second survey in Polvijärvi was carried out in 2001 (Joona & Ojanlatva 2002b) as part of the project "Polkuja esihistoriaan", which was organized by the North Karelian Museum. During the survey, ten new sites were found, and several loose-find places were located as well.

There have been some difficulties for the practical locating of sites as well as providing the z-coordinate; neither of which give the elevation above the Baltic Sea. Although

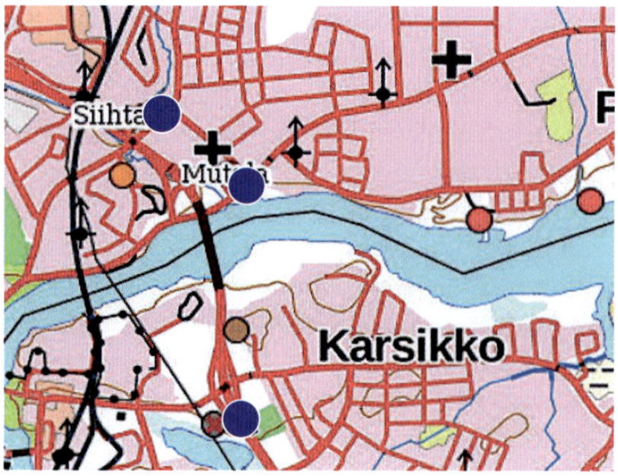

Figure 3. Mutala and Siihtala in Joensuu. The destroyed dwelling site of Varaslampi, the roundish, brown mark with red strokes, crossing each other on the northern side of Pond Varaslampi.

the elevations of the sites were later measured, the history of Lake Höytiäinen and the locations of the sites on the z-coordinate have sometimes been left out of the research. For this reason, only the relocating of Lake Höytiäinen is a process that is well-known, whereas the other parts of its history have lacked a detailed investigation. The same entails the history of Lake Viinijärvi. The situation with both Lake Pielinen and the northern parts of Lake Saimaa is better – including the Lakes Heposelkä and Pyhäselkä.

The first archaeological sites in Joensuu were found and studied by Sakari Pälsi (1937) and Jaakko Sauramo (1937) and after that in Mutala and by Ville Luho (1948) in Siihtala (Fig. 3). Pälsi excavated a Mesolithic stone hearth and Luho a Mesolithic dwelling site at Mutala, close to the city of Joensuu. They both date as early as 7700-7500 calBC (Joona & Ojanlatva 2002a: 8). Although the site had a very early dating, it included iron finds – an iron axe and iron slag. The first survey in Joensuu was carried out by Pirjo Lahtiperä (1968a) using 1:20 000 and 1:10 000 maps, after which the five dwelling sites dating to the Stone Age and one to the Iron Age were discovered. The next survey was done by the North Karelian Museum under the guidance of surveyors Juha-Pekka Joona and Eija Ojanlatva (2002a; 2002b).

The dwelling site of Varaslampi in Kettuvaara, first located already in the 1968 survey and carried out due to the building of the main road 6, was excavated by Maarit Suni (1974) and Lea Väkeväinen (1979). The excavation area was unusually large (3192 m^2), but the leaders of excavations stated that only a part of the dwelling site was researched. The site has been classified in the Finnish Register of Ancient Monuments and Sites (Muinaisjäännösrekisteri) as a Bronze Age site. The Textile ceramics and Luukonsaari Ware, the most common material at the site, however, connect it to the Early Metal Age culture.

Earlier surveys in Joensuu and its neighbouring municipalities were made already in the 1970s. Kiihtelysvaara, Tuupovaara, Eno and Pyhäselkä were finally incorporated into the municipality of Joensuu in the 2010s; these municipalities were also surveyed in the late 1970s. Kiihtelysvaara was surveyed by Oili Forsberg (2003a), and Eno by Petro Pesonen (2003a). Kiihtelysvaara and Tuupovaara were first surveyed by Pirjo Lahtiperä (1967a, 1967c) and many sites dated to the Historical Period. Oili Forsberg (2003a) and Taisto Karjalainen (2004) carried out the next large surveys on these municipalities in the project "In Ancient Paths". Some investigations were made after that, but before the municipality was joined to Joensuu. Pyhäselkä has been surveyed already by Pirjo Lahtiperä (1967b) but the larger fieldwork was made by Petro Pesonen (2003b) in the project "In Ancient Paths", after which 20 sites were made known.

In Kontiolahti, the first survey was made in 1968 (Lahtiperä 1968b) and was continued during the project "In Ancient Paths" (Forsberg 2003b). After that, some small areas in the municipality were surveyed by Mikroliitti LtD, for example in Jakokoski (Jussila 2012) and Kontioniemi (Jussila 2014). The copper mine in Kontiolahti was operational in the 18th century AD already (Nyman 2013).

The municipality of Liperi was first surveyed in 1966 when the number of sites was

19 (Sarvas 1966). Some sites in the area were investigated in the 1980s and '90s (Tuomi 1984) and the largest survey was carried out by Petro Pesonen in 2002 in the same project organized by the North Karelian Museum. During the survey, 36 new dwelling sites were discovered (Pesonen 2002: 7).

4 Dwelling sites in the area of four municipalities

In Table 1 can be seen the list of the number of dwelling sites of all periods that are known in the four municipalities in question. This Table is based on the information that has been collected from the Finnish Register of Ancient Monuments and Sites kept by the Finnish Heritage Agency.

The number of dwellings per site is the largest in Joensuu. This is because several municipalities around the city of Joensuu have been associated with it only after 2005 or later. They are the municipalities of Eno, Kiihtelysvaara, Tuupovaara, and finally Pyhäselkä in 2009, which enlarged the city's area. In other rural districts, the number of sites varies between 50 and 100. According to these observations, there are no clear concentrations of dwelling sites at any period of prehistoric time. Although some sites were inhabited for hundreds of years, they do not concentrate on the groups of sites. However, the sites indicate in detail how various the habitation itself has been. When looking at the material and their dating, it is possible to interpret the meanings of sites in the local areas.

Municipalities	number of dwelling sites	extent of municipalities (km²)
Polvijärvi	47	958,32 km²
Kontiolahti	41	1 029,83 km²
Joensuu	133	2 750,96 km²
Liperi	63	1 161,23 km²
Together	284	5 900,34 km²

Table 1. The number of dwelling sites in four municipalities in the province of North Karelia, in eastern Finland. The coverages of the municipalities in km² are presented in their situation at present.

The sites have primarily been found through surveys, and only secondarily in excavations and, finally, on the basis the of stray finds. In general, most of the archaeological sites in North Karelia date from the Stone Age, and the majority of them date to the Neolithic. The dating of the Stone Age dwelling sites is often based on the archaeological find material – quartz or flint scrapers and flakes – that are not necessarily easy to help in dating their period of use. Often, the Mesolithic dwelling sites are not counted as part of this group.

Table 2 will present the number of dwellings sites in all four municipalities. The data is based on the information calculated from the Finnish Register of Ancient Monuments and Sites. One must realize that, when calculating the number of dwelling sites in different phases, the results will usually be different. The main reason for this is the fact that new sites are found all the time and the

Municipalities	Stone Age	Early Metal Age / (Bronze Age)	Iron Age	Middle Age	Historical Period
Polvijärvi	36	1 (1)	1	0	16
Kontiolahti	37	2 (2)	0	0	1
Joensuu	109	2 (2)	0	0	21
Liperi	42	4 (1)	0	0	16
Together	224	9 (6)	1	0	54

Table 2. The number of dwelling sites in four municipalities in the North Karelia, in eastern Finland during the different period of the past.

Figure 4. Area 1, layer 4 at the dwelling site of Multavieru in 1997 from SE to NW. Foto: M. Lavento.

number is continually changing. The sites have been found in surveys, checking the find places of loose finds and observations carried out by amateur archaeologists. Furthermore, the sites have occasionally been reclassified with new datings of sites.

The results that are visible in the figures of dwelling sites are relatively simple (Figure 4). The Stone Age and Historical Period dwelling sites become evidently visible, whereas the other periods disappear from the others. The Iron Age and Middle Ages site are small in number, and they often remain devoid of information; at some sites, a Lapp cairn has been found (Saipio 2015: 14-15). This period becomes visible practically only at the multi-period sites.

The number of Early Metal Age dwelling sites is smaller than during the Neolithic Stone Age. Dividing the places into two different groups yields some inaccurate definitions. The known dwelling sites are slightly more in inland than in the coastal zone, although they have recently found more also there (Ikäheimo 2020).

Making the division of the hunter-gatherer sites between the Bronze Age and Early Metal Age groups needs good and transparent arguments. The main problem is that the same dwelling site can be classified on the basis of both categories. This problem makes the division imprecise and difficult to address with certainty. Although both periods began relatively synchronously, we know that the Early Metal Age has been dated in Finland proceeding to the end of Roman Iron Age, whereas the Bronze Age ends already in 500 calBC (Carpelan 1999: 271-273; Lavento 2018: 122-125). It is possible to see differences in the types of dwelling sites and how the populations have lived. But this difference is still problematic to decipher.

Iron Age dwelling sites are almost invisible in the area so far. In addition, sites from the Middle Ages and the Historical Period have been found in the area. One unusual site so far in Joensuu is Linnankukkula between Kiihtelysvaara and Tuupovaara on the eastern part of the city of Joensuu. It was found in 2016 and the site has been dated to between 1300 and 1400 AD (Laakso & Tiilikkala 2018).

5 The scientific analysis of ceramics at the dwelling site of Multavieru

Table 3 describes the Neolithic ceramics, divided by type, found in the different municipalities of North Karelia. It is also one means of following the activity and the change of each population in the areas.

The earliest ceramic type, the Säräisniemi 1 Ware, is known only at one site in the area. Although it has been dated in Finnmark in Norway during the period 5400-4500 calBC (Skandfer 2009: 357) or 5630-4100 calBC (Skandfer 2011: 171), it has been dated in Finland to ca 5200-4400 calBC (Pesonen & Leskinen 2009: 302). The Early Comb Ceramics is known in Polvijärvi and Kontiolahti at two sites. These early sites nowadays belonged to practically the same phase, and they date to the period ca 4300-3800 calBC (Nordqvist 2018: 58). They have recently been dated by the AMS to the shorter period, 4030-3830 calBC (Pesonen & Oinonen 2019: 256-261).

The Early Asbestos Ceramics is known in all of North Karelian´s municipalities and, therefore, in of special interest here. The different asbestos minerals were possible close to the sites, which began in ceramics during the early part of the 5th millennium BC (Lavento & Hornytzkyj 1996; Nordqvist 2018: 104-105; Pesonen 1996).

The Typical Comb Ware is known at 24 dwelling sites in the municipalities discussed in this article. During the time the ceramics were in use, the population grew to its largest extent during the Neolithic, although this is not evident on the basis of the available information at the moment (Mökkönen 2014; Tallavaara et al. 2014). The Typical Comb Ware people migrated to the area of contemporary Finland from the Middle- and Upper area of the Volga River. This took place soon after 3900-3800 calBC (Pesonen 2021: 28-36). The use of ceramics changed, when the Late Comb Ware, Kierikki Ware asbestos ceramics and even Rhomb-Pit ceramics came into use at some sites in Polvijärvi and Kontiolahti (Mökkönen & Nordqvist 2018). This indicated the various populations that prepared these ceramic types, which were synchronous for some time. The Rhomb-Pit ceramics is dated as slightly younger than the Late Comb Ware and it is known in Polvijärvi and Joensuu.

During the Middle and Late Neolithic, asbestos ceramics come into use, too. Asbestos ceramic types were first Early Asbestos Ware, then Kierikki Ware and finally Pöljä Ware. The Early Asbestos Ware came into use dur-

Ceramic types	Polvijärvi	Kontiolahti	Joensuu	Liperi	together
Pöljä Ware	2	1	4	13	20
Kierikki Ware	2			7	9
Corded Ware					
Pit-Comb Ware	1		2		3
Late Comb Ware	1	1			2
Typical Comb Ware	5	3	6	10	24
Early Comb Ware	1	1			2
Early Asbestos Ware	3	2	3	2	10
Säräisniemi 1 Ware			1		1
Together	15	8	16	32	71

Table 3. The number of Neolithic sites with separate ceramic types in different municipalities of North Karelia.

Municipalities	Polvijärvi	Kontiolahti	Joensuu	Liperi
Lepnaja Ware	7			
Iron Age ceramics	1			7
Kjelmøy Ware				
Sirnihta Ware	1			1
Luukonsaari Ware	1	3	4	3
Anttila Ware				
Kalmistonmäki Ware				
Kainuu Ware				
Tomitsa Ware	1		1	1
Sarsa Ware	1			

Table 4. Early Metal Age and Iron Age ceramic groups in the four municipalities of North Karelia.

ing the period 4600-3800 calBC (Nordqvist 2018: 58; Pesonen 1996). The Kierikki Ware came into use already ca 3600-2900 calBC (Nordqvist 2018) and Pöljä Ware was in use already during the period 3400-2500 calBC (Nordqvist 2018: 58; Nordqvist & Mökkönen 2017: j-k). Kierikki Ware is known in Polvijärvi and Liperi, but Pöljä Ware has been found in all four municipalities, which indicates when the use of asbestos ceramics reached a peak in ca 3400 calBC in North Karelia. Corded Ware has not been found in North Karelia.

One should note here that the Final Neolithic is a 500 year-long period, during which ceramic wares were not much found in the area during ca 2500-2000 calBC. The end of Pöljä Ware usage has usually been dated already at ca 2500 calBC. However, earlier in this period Jysmä Ware had come into use (Carpelan 1979: 15; Edgren 1964: 25). It must be remembered that we do not know enough about the end of the Final Neolithic. Recently, for example Nordqvist (2018: 107-108) has assumed that the type "does not exist as an individual type" but belongs to Pöljä ceramics.

The Early Metal Age sites imply a smaller number of fragments of ceramics than the Neolithic sites. The period is almost 2500 years shorter than the Neolithic. It begins in North Karelia with Textile ceramics, which has been divided in Finland into four subtypes (Lavento 2001). The Textile ceramics known in the area are mostly the eastern subtype of the main type.

Perhaps the most important dwelling site dating to the Early Metal Age is probably Varaslampi in Joensuu, which was excavated and then destroyed due to the construction of the main road 6. The material at the site includes mostly Textile ceramics but some Luukonsaari ceramics (Lavento 2001: 231).

In the area, Sär 2 ceramics was also in use. Its use began ca 800 calBC at the latest and it continued into the 4th century AD. In the Lake Saimaa region, two subtypes of Sär 2 ceramics were used in particular and two subtypes were both originally found and separated from the dwelling sites in Saimaa. The Luukonsaari type was originally decorated in Kuopio (Meinander 1969), but they were known in all municipalities in North Karelia. The Sirnihta Ware, which was originally separated in the municipality Kesälahti, in North Karelia (Carpelan 1999), is known still only at two dwelling sites in the municipalities of Polvijärvi and Liperi. Ceramics was at least locally in use at the end of the Early Metal Age, i.e. between ca 500 BC and 300 AD.

Of interest here is that the ceramics dating to the Iron Age ceramics and the Middle Ages was found at seven sites, but in different municipalities. The Iron Age ceramics are known in Liperi and the Lepnaja ceramics at several sites in Sotkuma in Polvijärvi (Enqvist 2002). The spinning of a potter's wheel began in the Karelian Isthmus in the mid-10th century BC and the using of the technology continued to the beginning of the 20th century (Uino 1997: 396-397).

6 The excavation at the dwelling site Multavieru in Kinahmo in Polvijärvi

Perhaps the best known archaeological site excavated in Polvijävi municipality is Sola in Suovaara, because of its early copper finds (Taavitsainen 1982). In the municipality of Polvijärvi a historical dwelling site was excavated here as well (Pesonen 1997a). However, the largest inhabited dwelling site from the Late Mesolithic to the Middle Ages was that of Multavieru in Kinahmo, in the northern part of the municipality.

Multavieru is the dwelling site area found within a zone of ca 350 m long and not larger than 80 m wide near the River Kiskonjoki, which empties into Lake Höytiäinen. The site was found in 1969 (Luukkainen 1969) by Tuomas Björkman (1969) and it was first investigated only after two years (Keskitalo 1971). The site was again surveyed by Hannu Poutiainen's group in 1993 and 1996 (Poutiainen et al. 1995; 1996). It was further studied at the end of the 1990s (Mikkola 1998; Pesonen 1997b) and finally in 2002 in connection with the survey of the whole of Polvijärvi (Joona & Ojanlatva 2002b).

Two excavations were carried out in Multavieru by the Department of Archaeology, the University of Helsinki in 1996 and 1997. The fieldwork was a teaching excavation and took place in the area where the site was partly destroyed, due to the operations of a gravel pit. Several periods of settlement were observed at the dwelling site. The excavated material implies material from the Late Mesolithic to the Middle Ages; this is visible in the change of stone material and the beginning of ceramic vessels and the change in the ceramic types in use at the site (Lavento 1997; 1999b: 30-32). The largest excavation area in Multavieru was 120 m^2 in 1996 (NM = National Museum number 29635:1-945) with its size being reduced to 113 m^2 in 1997 (NM 30767: 1-1018).

The excavation was led by Assistant Mika Lavento in 1996. The general maps were produced by Assistant FL Anne Wikkula together with Assistant BA Marja Mustakallio-Jussila. MA Derek Fewster was the expert who helped students to draw level- and profile mapping. The maps were drawn by BA Raija Ylönen and documented by MA Esa Mikkola.

Lavento led the excavation in 1997, but Lic. Phil. Petri Halinen was in charge of teaching general mapping and sampling. PhD Tuija Rankama, who was Acting Professor in Ari Siiriäinen's absence, taught drawing the level and profile mapping together with students. Esa Mikkola did the mapping of finds using a Tachymeter and drawing the maps together with three students.

The excavation was carried out in eight various areas, the largest in 1977 being 76 m^2 and the smallest 2 m^2 in size. One of them was the profile of the sandpit. The reason for this was that only part of the dwelling site was possible to excavate, whereas the area implied remains from different periods that were essential to understand. Two areas were excavated already in 1996 and the others in 1997. The four largest excavation areas and the profile are located close to the sandpit and the others to the north of it, the farthest being over 100 m away.

The research methodology at the excavation implied several natural scientific analyses that opened the viewpoint onto the long prehistory of the habitation at Multavieru. The first methods are based on the research of the geology of the environment that connected with the history of Lake Höytiäinen. The site was located near the shore of Lake Höytiäinen, until the water level drained down over 11,5 m in 1859 (Donner 1978, 228). We must remember that, before this, the water level was stable at almost the same level for several thousand years.

Area	flint point	flint artefact	flint flake	quartz artefact	quartz scraper	quart nuclei	quartz flake	stone fragment
1	1	1	8	3	28	23	1048	5
2					1	1	47	
3				1			95	
4			6	1				
5								
6							1	
rim of sandpit			2	1			27	

Table 5. Stone finds found in different excavation areas in Multavieru during 1996-1997 (NM 29635:1-945 and 30767:1-1018).

Area	ceramics rim	ceramics bottom	ceramics pieces	burnt clay	asbestos	burnt bones	iron slag
1	21	5	447	10 928	42	771	1
2			9	30			
3							
4	3		25	4		9	22
5						62	6
6							
rims of sand pits	1		1	514	42	125	1
together	26	5	483	11 476	42	967	30

Table 6. The amount of ceramic, asbestos, burnt bones and iron slag fragments in the excavations of Multavieru during 1996-1997 (NM 29635:1-945 and 30767:1-1018).

In Figure 5, one can see the location of the excavation areas of the field work during 1996-1997. In the excavations, several types of material were found to have been handled and prepared by human beings, but in some cases, these were difficult to date. Interestingly, though the site was in continuous use for a long time, there was only one flint point discovered, as well as one other artefact made from flint material. The 16 flint flakes indicate that very little raw flint material was brought to Multavieru. In the excavation finds, only a small number of other stone artefacts were found.

Table 6 indicates that a piece of burnt clay is the largest group of material finds at the dwelling sites in Multavieru. This was due to the historical iron casting at the site. The largest prehistoric materials at the site are burnt bones. An easily visible stone-find group is quartz and its flakes. This supports the assumption that they were preserved in use during the long period in the area. It is evident that, in spite of the long period of continuous living at the site, material was possible to find in the environment. Local quartz is the most important material for many artefacts.

Although ceramics is various and date to the long period of habitation, the number of pieces is very large. First of all, it was found in excavation area 1, which is the largest excavation area. Only evident ceramic pieces were separated in the excavations, together. Some ceramics were found particularly in areas 2 and 4 and the rim part of the site,

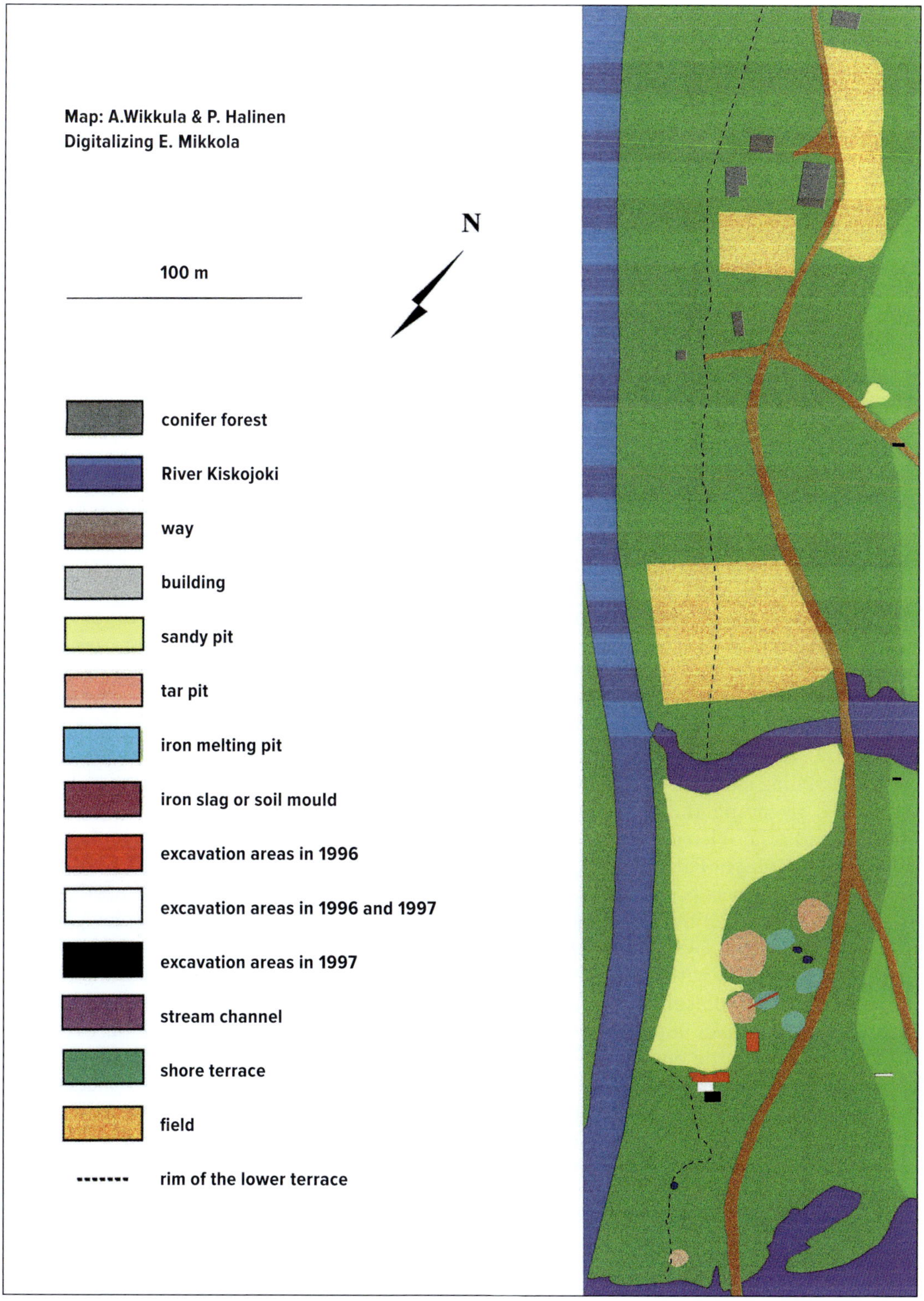

Figure 5. The general map of Multavieru teaching excavation area between 1996-1997.

which was mostly collapsed because of the gravel pit. The situation is the very same with other find types in the site with the ceramics. In other areas, ceramics were not found.

At the dwelling site of Multavieru, the ceramic material is rich and various and indicates how the site was inhabited during the whole Neolithic and Early Metal Age periods. The number of vessels is not large but indicates which ceramic types are visible at the site. The earliest groups at the site are Early, Typical and Late Comb Ware. The use of asbestos began with the Early Asbestos Ware but followed by Kierikki and Pöljä types. The habitation continued at the site during the Early Metal Age, which is easily visible in the Textile ceramics and Luukonsaari type of Sär 2 ceramics (Lavento 2001, 236).

What is particularly unusual is the appearance of Slavic-Karelian ceramics, which date to the Middle Ages. Ceramics came into use in northern Saimaa as early as ca 1000 AD and, interestingly, it continued as long as into the 19th century AD (Enqvist 2006: 412-413). In the area of Polvijärvi municipality, this ceramic has been found at several dwelling sites.

The dwelling site area in Multavieru was researched by doing the analysis of phosphorous and other elements. These analyses were done by student Paula Kouki. The topic of her MA thesis was based on the material at the site (Kouki 1999; 2000). The measurements were done by Atomic Absorption Spectroscopy (AAS) with FAAS and by Inductively coupled plasma mass spectrometry (ICP-MS). FAAS analysis belongs to AAS methods with the purpose of using the flame atomizer, the temperature of which is as high as 2300 degrees C. The method itself is the oldest, but the most commonly used in atomizing types in AAS flames (Broekaert 1998).

It became possible for Kouki (1999) to separate 11 elements, the amount of which was influenced by the activity of human beings. The elemental analysis indicated where and how active habitation had taken place in Multavieru. One problem – although a strength – for the soil analysis was the long settlement period from the Mesolithic to the Historical Period, which has mixed the earlier layers of soil. Although anomalies were possible for dating, the phases of the habitation were difficult to separate.

The dwelling site is located at the esker that developed when ice began to melt after the Dryas Age ca 14 000 BP. Soil at Multavieru is podsol, which has developed over the course of several hundred years and implies the aggregate sandy layers. Its main element after iron and aluminium are P and Ca, but Mg is connected with the high values of P/Ca. Zn and Mn have to accumulate with Ca and they are concentrated on the B horizon in ca 30 cm from the surface (Kouki 2000: 9-13). The analyses indicate the activity of human beings although the results are not self-evident.

Altogether, 289 phosphorous samples were taken altogether – 133 in 1996 and 156 in 1997. The analyses were carried out by the Arrhenius-test in the laboratory with Citron acid soaking (Jussila et al. 1989: 25). The values deverged quite a lot in the area, but the clustering is visible in two areas, in particular. The first concentration is located around the gravel pit, close to the excavations, and the second is located on the northern side of the small ditch and road in the research area, ca 150 m to the north of the excavation areas. The highest phosphorous values in the area were over 1000 mgP/kg and the smallest only 0 mgP/kg (Kouki 2000: 9-13; Lavento 1997: 12-13; 1999a: 17). This indicates active habitation that has lasted a thousand years at the same place and that the human activity has mostly concentrated to upper terrace.

Macrofossil analysis was done by BA Tanja Tenhunen with the help of FT Terttu Lempiäinen. Samples were taken from areas 1 and 4. The sites dating to the more recent part of the habitation, from Sär 2 ceramics and later, imply mostly rye (*Secele Cereale* and *Cereale*) which indicate grain crops. Fat hen (*Cheopodium Album*) and sorrel

Sample	area	year	Hel-number	^{14}C dating	calibrating
2		1996	3909	380±80	1449(0.57)1530 1560(0.43)1630
12		1996	3910	1480±80	450(0.14)490 510(0.86)650
13		1996	3911	5550±120	4590(003)4560 BC 4540(0.86)4320 BC 4290(011)4240 BC
10	rim	1997	4147	450±60	1400(1.0)1495
9	3	1997	4148	1950±80	90(1.0)180
4	3	1997	4164	1210±80	610(1.0)890

Table 7. Datings at the dwelling site of Multavieru in 1996 and 1997 calibrated by OxCal 4.1.

(*Rumex Acetosa*) in another sample indicated the collections from the Stone Age (Tenhunen 1988). In the 1996 analyses, the excavation mentioned that macrofossils are the most visible ones observed, but very visible are the raspberry remains (*Rubus Idaeus*), "the grape of the bear" (*Arctostaphylus Uva-ursa*), and common knotgrass (*Polygonum Aviculare*) (Tenhunen 1997). They are archaeophytes that have spread in the areas due to human impact.

FM Pirkko Ukkonen did the osteological analysis of the bone material. In general, the majority of the find material of the excavation was burnt bone and it was excavated in area 1. Bone remains were small but still possible to analyse. According to Ukkonen´s analysis, most of the bone material was fish and identified as *Teleostei*; the only exception she observed was Eurasian beaver (Lavento 1999a: 18). This is easy to understand because the place was located by the ancient shoreline and the excavations took place particularly there. The analysis is not done in detail from all the material of the excavations.

In 1996, 13 separate ^{14}C samples were taken and, in 1997, a total of 11 were taken. Together, six of these samples were analysed at the dating laboratory of Luomus, the Finnish Museum of Natural History, at the University of Helsinki in 1997 and 1999 (Table 7). Further, four TL samples were taken from the hearths in excavation area 3. So far, these samples have not yet been analysed.

Dating samples have come mainly from the different contexts of habitation, which were not easy to separate during the excavation, however. Two of these samples are quite recent and indicate that the dwelling site was used in the beginning of our era. The site was still inhabited during the early and middle Iron Age, which is unusual in North Karelia. This is from the Luukonsaari as well as Sirnihta ceramics found at 2/3 layers in the excavation areas and Slavo-Karelian ceramics in layers 1/2, in particular. It is probable that the most recent dates in Multavieru are probably between the 14th and 17th centuries AD (Lavento 1999b, 32-34) – though use of this site in Karelia has continued into the 19th century (Enqvist 2006, 417).

Multavieru is a versatile dwelling site where also a Historical Period is well visible. The first example of this period is located in excavation area 1 and on the SW side of the iron smelting places, which date between the 17th and the 18th centuries AD. Other remains of this period include tar burning that is visible in the area, as the tar burning pit was used in the 18th century AD. Both activities are well visible on the surface at the site (Lavento 1997). These activities were both used in the dwelling-site area because soil is mostly gravel and relatively easy to dig into. The wood that was used in both cases was pine.

Iron smelting was the method that was both practical and handy in eastern Finland during the Historical Period. Four historical smelting furnaces were located at the site (Fig. 5). One trench was opened across the smelting furnace in 1996. It indicates how the furnace was made and the similar type of structures it is possible to find in eastern Finland. They are known in the municipality of Ilomantsi in East Karelia, and they date between the 16th and 19th centuries AD (Mikkonen 2007) as well as to the same period as the smelting furnace in Multavieru. Iron ore was easy to hand quarry from Lake Höytiäinen and alongside the Kiskonjoki River and its tributaries. This gives an impression of how important the role of local iron smelting played for many historical local peasants around Lake Höytiäinen.

Three large tar-burning pits are visible inside the dwelling site, but probably their number was larger before digging the gravel pits. They are located outside the dwelling-site areas separated during the excavation project, too. This is because the same kind of area continues in the southern and northern part of it, and it is possible for that. The areas were utilized for both purposes and it means that the area was not needed for habitation. Numerous remains of tar burning were found, but the pits were not excavated during the fieldwork. Moreover, although the research of the historical tar trade has concentrated on Ostrobothnia in Finland (Turpeinen 2010), it is evident that large amounts of processing took place in the Saimaa region as well. In this regard, the archaeological research on tar is still awaiting work.

People who lived on the shores of Lake Höytiäinen began to seek suitable places for living that were well suited for agriculture; this took place at the beginning of the Early Metal Age, at the latest, but perhaps even earlier (Grönlund 1995; Grönlund et al. 1993; 1995). Multavieru was such a place; though the habitation continued there during the greater part of the Iron Age.

It has already been mentioned that the late Middle Ages and, in particular, the Historical Period starts becoming more visible at the dwelling sites also in North Karelia. One example of this was Kylälampi 2, which was excavated by Lic. Phil. Petro Pesonen and PhD Tuija Rankama in 1997. It was located on the western side of Multavieru, in Venepohja near Lake Viinijärvi (Pesonen 1997a). The habitation at the site began in the 1660s AD and it has continued there to the present. The research at the site was carried out by archaeological methods and natural scientific methods, such as macrofossil analysis. The excavation and the results achieved were interesting to the extent that they were new. The results were partly synchronous with the use of the dwelling site of Multavieru and its excavation was carried out in 1997. Further, the material at Kylälampi 2 implied the Slavo-Karelian ceramics, which was not well known.

7 Conclusions

The find material and sites constitute the totality of habitation from the late Mesolithic to the Historical Period in North Karelia. Almost all periods imply the archaeological remains of habitation. The area has been archaeologically researched by way of normal surveys; the extent of excavation work has been relatively restricted because the areas have not been actively built-up. Excavations have taken place mainly in cities and along some road alignments.

Most dwellings sites in North Karelia are known as a result of surveys. Few sites have been studied and only a small part of them have been excavated. Almost all ceramic types that can be assumed are known in the area. The information is visible in more detail in the chapter "Dwelling sites in the area of four municipalities". An unusually interesting ceramic group is that of the Slavic-Karelian Lepnaja group, which came into use in some areas during the Middle Ages.

At the end of the Stone Age, the habitation in these areas decreased considerably, but increased again during the Early Metal Age. This is visible at some dwelling sites where people lived during the late part of the

Early Metal Age. What the real situation was during the Iron Age, when iron came into use, is not known to us on the basis of contemporary archaeological material.

The population has been very small in general and the number of each group of people may have numbered only in some tens of people. The estimate for the largest population size possible, on the basis of the archaeological material, dates to the Typical Comb Ware II. This estimate is reflected in the numbers of dwelling sites and on the number of archaeological materials. The population receded at the end of the Final Neolithic but began to grow again from the beginning of the Early Metal Age. Habitation increased closer to our own era, but afterwards the early phase of the Iron Age was no longer so easily visible. This may refer to the receding habitation in the area. The settlement began again during the Viking Age and particularly toward the coming of the Middle Ages.

At the dwelling site of Multavieru in Polvijärvi municipality, a large part of the area was excavated and investigated by other methods. One must still remember that a considerable part of the dwelling site was destroyed as a result of digging a rather sizable gravel pit in the area, before the site was located. On the shores of Lake Höytiäinen, some areas were long in use at same places, because water or some raw materials were available at these places.

On the other hand, shore displacement in Lake Saimaa has affected the environment and, furthermore, it is changing all the time. The excavation material indicated the changes of the culture at the site, but it is essential that the habitation continues during the long term without gradually receding. Finally, the use of the Multavieru dwelling site was preserved during the Iron Age and the Middle Ages, when the area was inhabited by the farmers – who came to the northern Karelia area from the Karelian Isthmus, Häme and Savo. Thereafter, Multavieru was inhabited during the Middle- Ages and this habitation continued there into the Historical Age. The site was still used for other purposes, such as for making iron and finally tar burning during the 18th century AD.

References

Archival sources

Björkman, T. 1961. Kertomus Polvijärven pitäjän Solan kylän Suovaaran kivikautisen asuinpaikan kaivauksesta v. 1960.

Björkman, T. 1969. Kertomus kivikautisen asuinpaikan tarkastuksesta Polvijärven kunnan Lahto lahden kylässä Multavierun talon maalla 17.7.1969.

Forsberg, O. 2003a. Kiihtelysvaaran arkeologinen inventointi. Polkuja esihistoriaan -projekti. Pohjois-Karjalan museo.

Forsberg, O. 2003b. Kontiolahden arkeologinen inventointi. Polkuja esihistoriaan -projekti. Pohjois-Karjalan museo.

Grönlund, E., Koponen, M., Kupiainen, R. & Poutiainen, H. 2003. Tarkastus Polvijärven Multavierussa.

Joona, J.-P. & Ojanlatva, E. 2002a. Joensuun inventointi 13.5. – 31.5.2002.

Joona, J.-P. & Ojanlatva, E. 2002b. Polvijärven inventointi 1.6.-9.7.2002.

Jussila, T. 2012. Kontiolahti Jakokoski. Osayleiskaava-alueen muinaisjäännösinventointi 2012. Mikroliitti Oy.

Jussila, T. 2014. Kontiolahti Kontioniemen osayleiskaava-alueen muinaisjäännösinventointi 2014. Mikroliitti Oy.

Karjalainen, T. 2004. Tuupovaaran arkeologinen inventointi. Muinaisilla poluilla -hanke. Pohjois-Karjalan museo.

Keskitalo, O. 1971. Polvijärvi Martonvaara Multavieru. Kivikautinen asuinpaikka.

Koivikko, M. and Ruuttu, P. 1996. Polvijärvi Sola Suovaara prospektointi.

Laakso, V. & Tiilikkala, J. 2018. Joensuu Kovero. Linnankukkula Muinaislinnan arkeologinen kaivaus 2018. Muinaislinnan arkeologinen kaivaus 2018. Maanala Oy.

Lahtiperä, P. 1967a. Kiihtelysvaaran arkeologinen inventointi. Muinaistieteellinen toimikunta.

Lahtiperä, P. 1967b. Pyhäselän arkeologinen inventointi. Muinaistieteellinen toimikunta.

Lahtiperä, P. 1967c. Tuupovaaran arkeologinen inventointi. Muinaistieteellinen toimikunta.

Lahtiperä, P. 1968a. Joensuun arkeologinen inventointi. Muinaistieteellinen toimikunta.

Lahtiperä, P. 1968b. Kontiolahden arkeologinen inventointi. Muinaistieteellinen toimikunta.

Lavento, M. 1997. Polvijärvi Kinahmo Multavieru. Kertomus v:n 1996 kaivauksesta. Helsingin yliopisto,

arkeologian laitos.
Lavento, M. 1999a. Polvijärvi Kinahmo Multavieru. Kertomus v:n 1997 kaivauksesta. Helsingin yliopisto, arkeologian laitos.
Luho, V. 1948. Kertomus Pielisjoen suun Siihtalan kivikautisen asuinpaikan tutkimuksesta 1.6. – 4.6.48.
Meinander, C. F. 1962. Kertomus intendentti C. F. Meinanderin kaivauksista Polvijärven pitäjän Solan kylässä sijaitsevalla Suovaaran kivikautisella asuinpaikalla 27. - 29. 7. 1961.
Mikkola, E. 1999. Polvijärvi [6] Multavieru. Kivikautisen asuinpaikan tarkastus.
Mikkola, E., Vuoristo, K., Enqvist, J. & Koivikko, M. 1998. Polvijärvi [1] Suovaara. Kivikautisen asuinpaikan tarkastus.
Nyman, J.-E. 2013. Kontiolahti, Hokka. Arkeologinen tarkastus historiallisen ajan kaivoskohteella. Museovirasto.
Pälsi, S. 1937. Pielisensuun Mutalan kivikautinen liesi.
Pesonen, P. 1997b. Polvijärvi [6] Multavieru. Kivikautisen asuinpaikan tarkastus.
Pesonen, P. 2002. Liperin arkeologinen inventointi. Pohjois-Karjalan museo.
Pesonen, P. 2003a. Enon arkeologinen inventointi. Pohjois-Karjalan museo.
Pesonen, P. 2003b. Pyhäselän arkeologinen inventointi. Pohjois-Karjalan museo.
Poutiainen, H., Koponen, M. and Kupiainen, R., 1996. Pohjois-Karjalan metallikautisen ja keski-aikaisen asutuksen avainalueiden arkeologinen inventointi 1992-1994.
Sarvas, P. 1965. Polvijärven kiinteät muinaisjäännökset. Polvijärven kunnan kiinteiden muinaisjäännösten inventointi kesällä 1965. Muinaistieteellinen toimikunta.
Sarvas, P. 1966. Liperin pitäjän kiinteän muinaisjäännökset. Muinaistieteellinen toimikunta.
Suni, M. 1974. Joensuu Varaslampi. Pronssikautisen asuinpaikan kaivaus.
Tenhunen, T. 1998. Mikrofossiilianalyysi Polvijärvi Multavieru 1996. Polvijärvi Kinahmo Multavieru. Kertomus v:n 1997 kaivauksesta.
Väkeväinen, L. 1979. Joensuun Varaslampi. Varhaismetallikautisen asuinpaikan koekaivaus.

Literature

Anttila, V. 1967. *Järvenlaskuyhtiöt Suomessa*. Kansatieteellinen Arkisto 19.
Broekaert, J. A. C. 1998. *Analytical Atomic Spectrometry with Flames and Plasmas*, 3rd Edition, Wiley-VCH, Weinheim, Germany.
Carpelan, C. 1979. Om asbestkeramikens historia i Fennoskandien. *Finskt Museum* 1978: 5–25.
Carpelan, C. 1999. Käännekohtia Suomen esihistoriassa aikavälillä 5100...1000 eKr. In P. Fogelberg (ed.), *Pohjan poluilla. Suomalaisten juuret nykytutkimuksen mukaan*. Suomen Tiedeseura, Helsinki: 249–279.
Donner, J. 1978. *Suomen kvartäärigeologia*. Geologian laitos, Geologian ja paleontologian osasto. Moniste N:o 1. Helsingin yliopisto.
Edgren, T. 1964. Jysmä i Iidensalmi. *Finskt Museum* 71: 13-37.
Ehlers, C., Lindroos, A. & Selonen, O. 1993. The late Svecofennian granite-migmatite zone of southern Finland – a belt of transpressive deformation and granite emplacement. *Precambrian Research* 64: 295-309.
Enqvist, J. 2002. Polvijärven Sotkuman karjalainen keramiikka. *SKAS* 4/2002: 26-34.
Enqvist, J. 2006. Karelian Ware: pottery of Slavonic type in eastern Finland. In V.-P. Herva (ed.): *People, Material Culture and Environment in the North. Proceedings of the 22nd Nordic Archaeological Conference, University of Oulu, 18-23 August 2004*. Studia humaniora ouluensia 1: 412-419.
Grönlund, E. 1995. A Palaeoecological study of landuse history in East Finland. *Joensuun yliopiston luonnontieteellisiä tutkimuksia* No: 31.
Huurre, M. 1985. Joensuun alueen esihistoria. *Joensuun kaupungin historia* 1 (Toim. Kalevi Ahonen et al.): 27-55.
Ikäheimo, J. 2020. From Obvious to Ambiguous: a Comparative Case Study of Crucible Fragments from a Bronze Age Site in northern Finland. *Fennoscandia archaeologica* XXXVII: 25-44.
Jussila, T., Lavento, M., Schulz, H.-P. 1989. *Maaperän fosforianalyysi arkeologiassa*. Helsinki Papers in Archaeology No. 3.
Kirkinen, H. 1963. *Karjala idän kulttuuripiirissä. Bysantin ja Venäjän yhteyksistä keskiajan Karjalaan*. Helsinki: Suomen historiallinen seura.
Koistinen, T. J. 1981. Structural evolution of an early Proterozoic strata-bound Cu-Co-Zn deposit, Outokumpu, Finland. *Transactions of the Royal Society of Edinburgh: Earth Sciences* 72: 115-158.
Kouki, P. 1999. *Polvijärven Multavierun maaperäkemialliset analyysit*. MA-thesis. Manuscript at the Institute for Cultural Studies, Department of Archaeology, University of Helsinki.
Kouki, P. 2000. Polvijärven Multavierun asuinpaikan geokemialliset analyysit. *Muinaistutkija* 3/2000: 2-14.
Lavento, M. 1999b. Statistical approach to archaeological stratigraphy of a multi-period dwelling site at Multavieru in Polvijärvi. In M. Huurre and the editorial board, C. Carpelan, P. Halinen, T. Kirkinen, V. Laulumaa, M. Lavento and M. Lönnqvist (eds.): *Dig it all, papers dedicated to Ari Siiriäinen*: 29-38. Helsinki: The Finnish Antiquarian Society, The Archaeological Society of Finland.
Lavento, M. 2001. *Textile ceramics in Finland and on the Karelian Isthmus: nine variations and fugue on a theme of C. F. Meinander*. Suomen Muinaismuistoyhdistyksen Aikakauskirja 109. Helsinki: Suomen Muinaismuistoyhdistys.
Lavento, M. 2018. Pronssin valajat (1900 eKr. – 400 jaa.). In A. Talka & J. Paaskoski (eds.): Rajamaa. *Etelä-Karjalan historia* I: 121-165.
Lavento, M. & Hornytzkyj, S. 1996. Asbestos types and their distribution in the Neolithic, Early Metal Period and Iron Age Pottery in Finland and Eastern Karelia.

Pithouses and Potmakers in Eastern Finland. Helsinki Papers in Archaeology 9: 41-70.

Meinander, C.F. 1969. Dåvits. En essä om förromerskt järnåldern. *Finskt Museum* 76: 27-69.

Mikkonen, T. 2007. Järvimalmin jalostus Ilomantsissa. *Tekniikan Waiheita* 2007/4: 17-20.

Mökkönen, T. 2014. Archaeological Radiocarbon Dates as a Population Proxy. A Spectical View. *Fennoscandia archaeologica* XXXI: 125-134.

Mökkönen, T. & Nordqvist, K. 2018. Kierikki Ware and the contemporary Neolithic asbestos- and organic-tempered potteries in north-east Europe. *Fennoscandia archaeologica* XXXIV (2017): 83–116.

Nordqvist, K. 2018. *The Stone Age of Northeastern Europe 5500–1800 calBC. Bridging the gap between the East and the West*. Acta Universitatis Ouluensis B Humaniora 160.

Nordqvist, K. & Herva, V.-P. 2013. Copper Use, Cultural Change and Neolithization in North-Eastern Europe (c. 5500-1800 BC). *European Journal of Archaeology* 16 (3) 2013: 401-432.

Nordqvist, K. & Mökkönen, T. 2017. Periodisation of the Neolithic and radiocarbon chronology of the Early Neolithic and the beginning of Middle Neolithic in Finland. *Documenta Praehistorica* 44: 78–87.

Oinonen, M., Pesonen, P., Alenius, T. Heyd, H., Holmqvist-Saukkonen, E. Kivimäki, S., Nygrén, T., Sundell, T. & Onkamo, P. 2014. Event reconstruction through Bayesian chronology: Massive mid-Holocene lake-burst triggered large-scale ecological and cultural change. *Holocene* 2014: 1-9.

Pesonen, P. 1996. Early Asbestos Ware. In T. Kirkinen & M. Tusa (eds.): *Pithouses and potmakers in eastern Finland: reports of the Ancient Lake Saimaa project*. Helsinki papers in archaeology no. 9. Helsinki: Helsingin yliopisto: 9–39.

Pesonen, P. 1997b. Karjalaista keramiikka Polvijärven Sotkumasta. Historiallisen ajan asuinpaikan kaivaustuloksia. In Ranta, H. (ed.): *Kentältä poimittua* 4, kirjoitelmia arkeologian alalta. Museoviraston arkeologian osaston julkaisu n:o 7: 75-87.

Pesonen, P. 2021. *Continuity and Discontinuity in Early, Middle and Late Neolithic Pottery Types of Eastern Fennoscandia. Reflections from Bayesian Chronologies*. Doctoral Thesis. Helsinki University.

Pesonen, P. & Leskinen, S. 2009. Pottery of the Stone Age huntergatherers in Finland. In P. Jordan & M. Zvelebil (eds.): *Ceramics before farming: the dispersal of pottery among prehistoric Eurasian huntergatherers*. Walnut Creek: Left Coast Press: 299–318.

Pesonen; P. & Oinonen, M. 2019. The chronology of Jäkärlä Ware: Bayenian Interpretation of the Old and the New Radiocarbon Dates from Early and Middle Neolithic southwest Finland. *Documenta Praehistorica* XLVI: 246-267.

Pihlatie, M., Vesajoki, H., Lappalainen, A., Hottola, P. 1997. *Höytiäinen = Lake Höytiäinen*. Artaround.

Poutiainen, H., Grönlund, E. & Koponen, M. 1995. Joining the Forces: archaeologists and palaeo-ecologists in the traces of North Karelian settlement and land-use History. *Fennoscandia archaeologica* XII: 153-159.

Pälsi, S. 1937. 1. Kaivauskertomus. Pielisensuun Mutalan kivikautinen liesi. *Suomen Museo* XLIV: 1-6.

Saarnisto, M. 1968. The Flandrian history of Lake Höytiäinen, Eastern Finland. *Bull. Geol. Soc. Finland* 40: 71-98.

Saipio, J. 2015. Nastola Kilpisaari 2 – Esihistoriallisen röykkiön elämänkaari. *Muinaistutkija* 1/2015: 2–18.

Sauramo, M. 1937. Pielisensuun Mutalan kivikautinen liesi. II Löytöpaikan geologia. *Suomen Museo* XLIV: 6-13.

Sauramo, M. & Auer, V. 1928. On the development of the Höytiäinen in Carelia and its ancient flora, *Comm. Inst. Quaest. Forest. Finlandiae* 13: 1-54.

Skandfer, M. 2009. 'All change'? Exploring the Role of Technological Choice in the Early Northern Comb Ware of Finnmark, Arctic Norway. In P. Jordan and M. Zvelebil (eds.) *Ceramics Before Farming. The Dispersal of Pottery Among Prehistoric Eurasian Hunter-Gatherers*. Walnut Creek: Left Coast Press: 347–374.

Skandfer, M. 2011. Early pottery in northern Fennoscandia. In Hartz, S., Lüth, F. & Terberger, T. (eds.): Frühe Keramik in Ostseeraum – Datierung und Sozialer Kontext. Internationaler Workshop in Schleswig vom 20. bis 21. Oktober 2006. *Bericht der Römisch-Germanischen Kommission* Band 89 (2008): 165–180.

Taavitsainen, J.-P. 1982. A Copper Ring from Suovaara in Polvijärvi, Northern Karelia. *Fenoscandia archaeologica* I: 41-49.

Tallavaara, M., Pesonen, P., Oinonen, M. & Seppä, H. 2014. The Mere Possibility of Biases does not invalidate archaeological Population Proxies – Response to Teemu Mökkönen. *Fennoscandia archaeologica* XXXI: 135-140.

Tenhunen, T. 1997. *Kasvijäänteet arkeologiassa. Polvijärven Multavierun makrofossiilitutkimus*. Proseminaarityö Helsingin yliopiston arkeologian laitoksella.

Tuomi, M.-L. 1984. *Suur-Liperin historia*. Joensuu.

Turpeinen, O. 2010. *Mustan kullan maa: Tervan historia*. Somero: Oy Amanita Ltd.

Uino, P. 1997. *Ancient Karelia – Archaeological studies. Muinais-Karjala – Arkeologisia tutkimuksia*. Suomen Muinaismuistoyhdistyksen Aikakauskirja 104. Helsinki: Suomen Muinaismuistoyhdistys.

Archaeological Survey of the Outer Islands of the Gulf of Finland in 2019

Maria A. Razzak (Yushkova)

Maria A. Razzak, Institute for the History of Material Culture Russian Academy of Sciences. Address: 191186, Saint-Petersburg, Dvortzovaya embankment, house 18, letter A, e-mail: marija_ju@mail.ru

Abstract

In 2019, an expedition of the Institute for the History of Material Culture (IIMK) RAS conducted archaeological surveys on the following islands in the archipelago of Vneshniye Ostrova (Outer Islands, *Finnish*: Suomenlahden ulkosaaret) in the Gulf of Finland: Moshchny Island (Lavansaari), Maly (Peninsaari), Seskar (Seiskari) and Yuzhny Virgin (Itä-Viiri). Familiar sites were re-examined and new ones were revealed. On Moshchny Island, a cairn, a burial ground, a Bronze Age – Early Iron Age settlement, a stone foundation of a church of the Modern Period and a group of stone heaps were discovered. On Yuzhny Virgin island, stone structures including labyrinths, circles and heaps of stone were examined. The studied sites are culturally close to the antiquities of the northern littoral of the Gulf of Finland (the Vyborg region of Leningrad oblast and southeastern Finland).

1 Introduction

In 2019, an expedition of the Institute for the History of Material Culture (IIMK) RAS conducted archaeological surveys in the archipelago of the Outer Islands (Suomenlahden ulkosaaret) in the Gulf of Finland. These studies were part of an interdisciplinary scientific investigation organised by the Leningrad Regional Branch of the Russian Geographical Society. The work was carried out on the islands of Moshchny (Lavansaari), Maly (Peninsaari), Seskar (Seiskari) and Yuzhny Virgin (Itä-Viiri) that belong to the Vneshniye Ostrova islands in the Gulf of Finland[1] (Fig. 1). In total, three trips, each lasting about a month, were conducted; most of the time was devoted to surveys on Moshchny Island (Lavansaari).

The history of archaeological investigation of the Outer Islands comprises two periods. The first is the pre-war period, when the islands were part of Finland. During those years, information from local residents about particular archaeological sites and finds from the islands was accumulated and professional archaeological excavations and surveys were carried out. Afterwards, during the second half of the 20th century, these islands were unavailable for scientific explorations on account of being within the Russian border zone. During this period, Finnish researchers summarised and analysed the materials that had been already accumulated. Information on archaeological sites of the Outer Islands was collected and studied by T. Miettinen (1996) and summarised in a monograph by P. Uino (1997). A separate investigation by T. Edgren (1993) was dedicated to analysis of the results of excavations conducted by S. Pälsi at the cemetery of Suursuonmäki on Moshchny Island. In an article by C. Carpelan and P. Uino (2003), the available information about the finds and excavations of the first half of the 20th century on the island of Bolshoy Tyuters (Tytärsaari) was summarised. A. I. Saksa considered the available data on antiquities of the islands

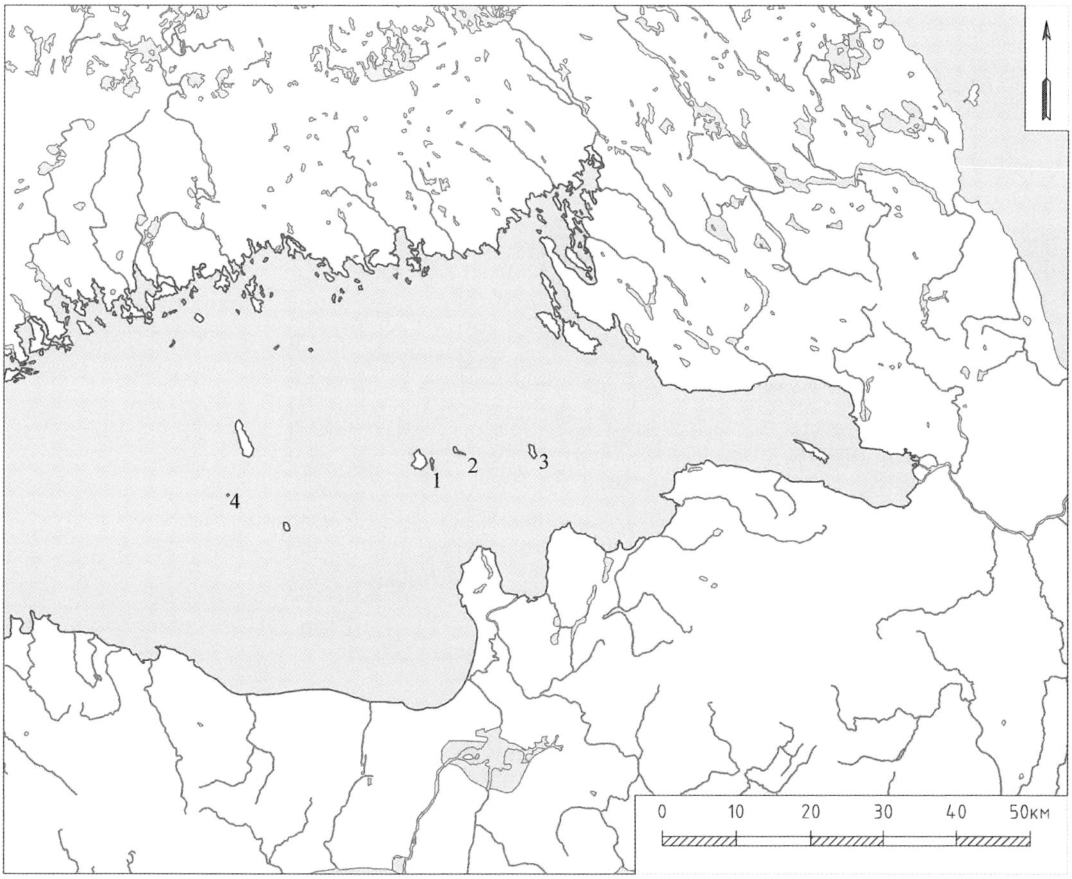

Figure 1. Islands of the Gulf of Finland where the archaeological surveys of 2019 were conducted. 1 – Moshchny (Lavansaari), 2 – Maly (Peninsaari), 3 – Seskar (Seiskari), 4 – Yuzhny Virgin (Itä-Viiri).

in the context of the evolution of archaeological cultures of the northern and southern coasts of the Gulf of Finland (Saksa 2014).

A new period of field studies on the islands began in 1999, when archaeological surveys were conducted by A. I. Saksa on Moshchny Island (1999), by G. S. Lebedev on Hogland (Suursaari) (2000), and by E. K. Blokhin on Hogland (2013–2015). However, the results of these investigations have not been published.

2 Field investigation of 2019

Moshchny Island. By the beginning of the surveying explorations on the island, the following sites were known: 1. burial ground of Suursuonmäki with stone burial cairns where Sakari Pälsi carried out excavations in 1930; 2. stone burial cairn and stone circle known from 1926 onwards through information given by local teacher Johannes Suomalainen; 3. group of stone cairns revealed by A. I. Saksa in 1999. The investigations were started with the search for and examination of the known sites; after their re-discovery, surveys were conducted on the rest of the island. The efforts in 2019 on Moshchny Island resulted in the uncovering of sites dating from the Bronze Age to the Modern Period (Fig. 2).

Stone burial cairn of Suisaari. The promontory of Suisaari, separated by a sandy neck of land, is situated on the eastern section of

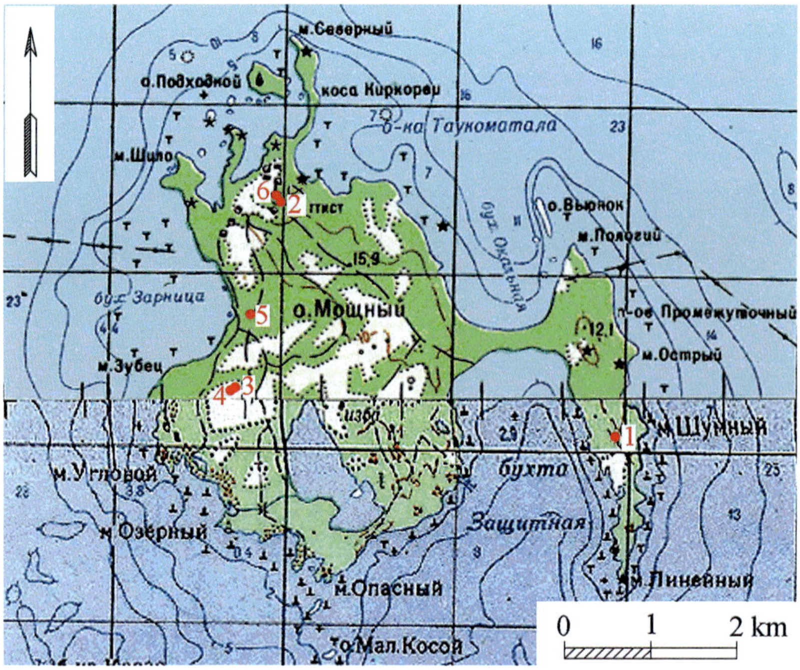

Figure 2. Map of Moshchny Island showing the revealed sites. 1 – burial cairn of Suisaari, 2 – settlement of Lavansaari, 3 – the cemetery of Suursuonmäki, 4 – burial cairn Suursuonmäki 2, 5 – a group of three cairns, 6 – a stone foundation of a church.

Moshchny Island. According to information by local teacher J. Suomalainen in 1926, a stone cairn 10 m in diameter and about 1.5 m high was located on the Suisaari Peninsula. In addition, J. Suomalainen made a note about the finding of a stone circle ca 7.5 m in diameter not far from the cairn (Edgren 1993; Uino 1997: 269-270).

During a survey of the peninsula, a stone cairn was found in the forest on the southern promontory of the upper terrace. It was situated on the southern spur of a ridge extended meridionally in the central part of the Suisaari Peninsula. The diameter of the cairn was 12–13 m, and its height was up to 1.4 m; it was constructed of waterworn boulders covered with turf (Fig. 3). The presumed height above sea level was about 8 m.

A group of cairns[2] on the northern coast of the Gulf of Finland west of Vyborg, near the village of Bolshoy Bor, was recently discovered and examined. This group is situated on a rock outcrop overlooking the sea 9.5–10.5 m above sea level. Four stone burial cairns were investigated; accumulations of calcined bones, pieces of charcoal and a bronze knife were discovered beneath the mounds. It can be theorised, based on a combination of the topography, size and shape of the cairns, that these are Bronze Age cairns (*hiidenkiuas*), which are characteristic of the northern littoral of the Gulf of Finland. The eastern boundary of the distribution of cairns of this type runs over the northern coast of the Gulf of Finland (Meinander 1954: 116–117; Miettinen 2012; Lavento 2015). The investigated cairns were dated to between the end of the 2nd and the beginning of the 1st millennium BC (period IV, i.e. ca 1100–900 BC or slightly later). As the Suisaari cairn has a similar topography, size, and shape, we presume that it dates from the same period. However, a later dating cannot be completely ruled out since it is known from the example of excavations of cairns in Finland that a significant part of them belong to the Iron Age (Tuovinen 2002: 66; Asplund 2008: 73).

Settlement of Lavansaari. The settlement-site of Lavansaari was first discovered during the present investigations. It is situated in the northern section of the island on the site of an abandoned Finnish village. An exploratory trench was started with an area of 16 sq. m. (Fig. 4) 15 m to the east from the stone fence of the church, near the road crossing. In this trench, strata from different periods were revealed. Finds dated to the existence of the Finnish village from the 19th century to the first half of the 20th century were retrieved from the upper layer of grey-brown sandy loam – fragmentary bricks, iron nails and cramps, fragments of porcelain and clay ware.

Below, under a layer of grey sandy loam devoid of finds, a surface of yellow virgin sand was cleared and two complexes were investigated. One complex represented a hearth; it was located in a pit up to 0.4 m deep and was faced with stones over the bottom; in its fill, charcoal and dark-grey sand intercalations were detected (Fig. 5). The second complex was a spot of the cultural layer of a grey-yellow colour measuring about 1,4 × 2 m in area and up to 0,15 m thick. These layers were about 6 m above sea level.

In the course of the excavation of the spot of cultural layer and clearing of the hearth, fragments of ceramic pottery with a hatched surface were found. In total, 42 fragments were retrieved, possibly belonging to a single vessel with fragments of walls and bottom parts preserved (Fig. 6). The thickness of the vessel's walls was 0,7–1 cm; the clay paste contains a mineral temper of rotten stone; the sherds are brown in colour and their state of preservation is poor (exfoliation). The main peculiarity of the ceramic fragments is in the treatment of their outer surface. The latter is covered with strokes; these are mostly vertically arranged; the inner surface of the sherds is smooth. Possibly, the strokes were produced with a bunch of grass. It is difficult to reconstruct the shape of the vessel;

Figure 3. Burial cairn of Suisaari. Photo: M. Razzak, 2019.

Figure 4. Settlement of Lavansaari. The location of the exploratory trench. Photo: M. Razzak, 2019.

Figure 5. Settlement of Lavansaari. The hearth in the process of excavation. Photo: M. Razzak, 2019.

the preserved walls show a straight profile, with the pot having had a flat bottom with a diameter of about 14 cm.

The closest parallels to the above pot are represented by ceramic vessels found on the island of Bolshoy Tyuters. Here, fragments of pottery with a hatched surface were found during excavations in the dunes. At least two vessels can be reconstructed; one is well-preserved and shows a flat bottom, practically straight walls and a gently curving S-shaped profile of the upper body. This type of pottery possibly belongs to the Paimio type, which generally is datable to within 1500–700 BC or to a younger group of the Morby ware of the Early Iron Age (Carpelan & Uino 2003: 83).

During the excavations, it was presumed that the hearth and the spot of cultural layer with the pottery finds are synchronous; samples for radiocarbon analysis were selected from the hearth and charcoal intercalations. The analysis yielded seven dates indicating three differing chronological horizons. Three dates – 1010–909 calBC (2808±40 SPb-3062), 974–834 calBC (2765±40 SPb-3087) and 930–829 calBC (2745±50 SPb-3088) – are close to each other and after calibration with 68,3% probability suggest a time period between the 10th and 9th centuries BC.

Figure 6. Settlement of Lavansaari. Stroked pottery. Photo: I. Samovarov, 2019.

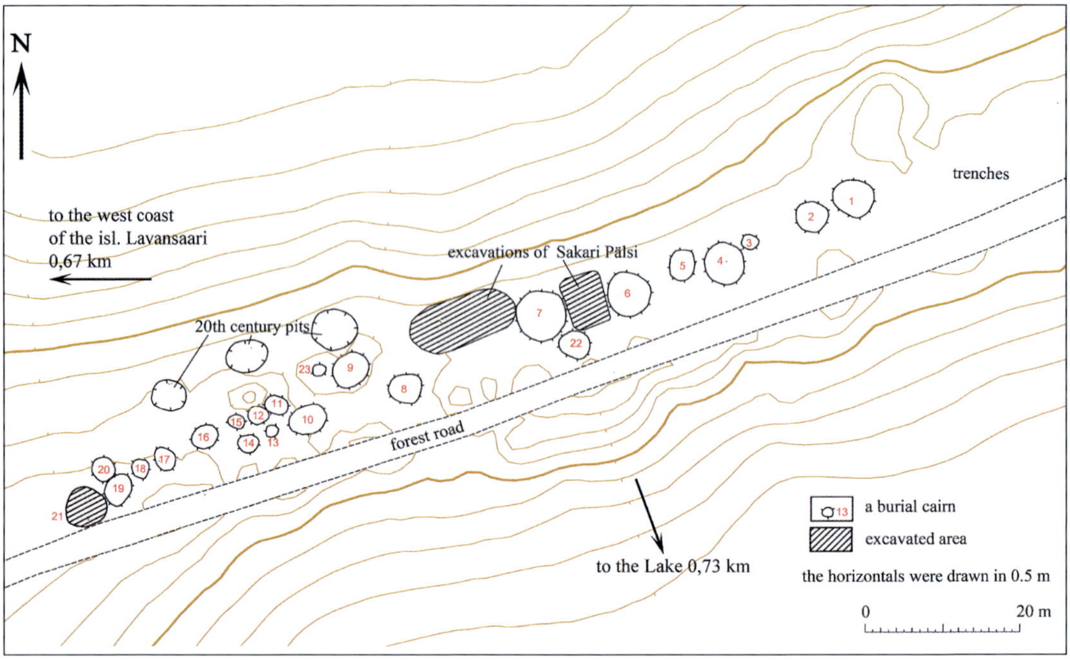

Figure 7. Topographic plan of the cemetery of Suursuonmäki. 2019

These dates were gained from the fragments of charcoal extracted from the spot of the cultural layer (samples SPb-3087, 3088) and from the hearth (sample SPb-3062). Another date obtained from the cultural layer corresponds to the pre-Roman Iron Age, i.e. 355-164 calBC (2175±40 SPb-3065). The other group of three dates from samples from the hearth indicates the second half of the 13th century AD[3].

Another AMS date is obtained from the soot on one of the ceramic fragments: 161-55 calBC (2102±30 Hela-4803). As the soot sample on the pottery dates it more accurately than the charcoal from the cultural layer, this date can be considered the most reliable and the pottery can be attributed to the pre-Roman Iron Age. However, the presence of two other date ranges from the cultural layer and the hearth, with one belonging to the Bronze Age and the other to the medieval period, leaves open the question about the dating of the settlement.

The cemetery of Suursuonmäki. This cemetery was discovered by local teacher J. Suomalainen in 1926. J. Suomalainen noted about the finding of 32 cairns. In 1930, investigations at the burial ground were conducted by S. Pälsi. He drew a plan of the cemetery, marking 28 cairns and excavating three of the mounds. Most of the cairns are of an approximately round plan with a diameter of about 3–4 m; one cairn was of an elongated form measuring ca 11 × 4 m. S. Pälsi carried out an excavation of this elongated cairn and of two other cairns with the roughly circular plan. Calcined human bones, ashes and charcoal, and pieces of resin were discovered in the cairns. In 1993, T. Edgren published an article where he described in detail and analysed the results of the excavations of this burial ground. Through samples from the excavations by S. Pälsi three radiocarbon dates were obtained, which, after calibration, indicate the last centuries BC – first century AD: 375 – 52 calBC (2165±60 BP Ua-2547), 151 calBC – 219 calAD (1975±70 BP Ua-2545) and 107 calBC – 236 calAD (1960±70 BP Ua-2546), calibrated with OxCal v4.4.2. In 1999, the burial site was explored by A. I. Saksa.

Figure 8. Stone foundation of the church. Photo: M. Razzak, 2019.

The cemetery under study is situated in the western section of the island 0.7 km from the seashore, in a forest on a ridge elevated 4 m above the surrounding woodland. A visual examination revealed 23 cairns 1,5–6 m in diameter and up to 0,5 m high (Fig. 7). In addition, the locations of the two cairns excavated in 1930 were found. The cairns were laid from stones, the latter now covered with turf. The plan of the mounds is round-oval; the tops of most of the cairns are flattened. The cemetery is about 110 m long and 12 m wide. The cairns extend in a chain along a small height. The mounds are distanced 0,5-6 m from each other. A separately located cairn named Suursuonmäki 2 was revealed about 130 m westward from the burial ground. Constructed of stones, it has a diameter of ca 5,5 m and a height of up to 0,5 m. Its top is flattened and the surface covered with turf. The presence of two cairns to the west of the cemetery was mentioned by S. Pälsi. It is possible that it was one of these cairns that was thus found recently.

The closest parallel to the Suursuonmäki cairns is represented by the cemetery of Pyhtää Strukankalliot west of the city of Kotka. The cemetery consists of 10 stone cairns of similar dimensions, of which three were excavated. In the mounds, objects of the early Roman period were found, viz. fibulae, gold-glass beads, etc. (Miettinen 1997).

In addition, on Moshchny Island a group of three stone piles was revealed. This group was discovered in 1999 by A. I. Saksa in the western part of the island at the edge of a dune. According to A. I. Saksa, the definition of sites of this type varies widely and may be dated to the Middle Ages or the Modern Period.

A stone foundation of a church is preserved in the central area of the abandoned Finnish village near the settlement of Lavansaari (Fig. 8). This wooden church on a stone base was built in 1783 and destroyed during World War II. Its general dimensions are ca 18 × 23 m; the foundation is constructed of granite blocks; the width of the foundation is 0,35–0,6 m. The church is sur-

Figure 9. Yuzhny Virgin southern labyrinth. Photo: M. Razzak, 2019.

Figure 10. Yuzhny Virgin northern labyrinth. Photo: M. Razzak, 2019.

rounded with a boulder fence and is octagonal in plan.

Seskar. At the southern extremity of the dune ridge extending meridionally along the entire island, a small fragment of pottery and a quartz flake were found. The ceramic fragment possibly belongs to a handmade vessel and can be dated widely to within a range from the early Iron Age to the Middle Ages. No new finds were discovered around its findspot. Surveys on Maly revealed no new sites.

Yuzhny Virgin. This is a rather small island with general dimensions of about 200 × 600 m, practically devoid of vegetation. Most of the island is constituted by terraces of stone pebbles. It is situated 10 km eastward from Hogland.

As early as the first half of the 19th century, the academician Karl Ernst von Baer, during his scientific expedition in the Gulf of Finland, made an unplanned visit to the island of Y. Virgin. He wrote a description of the island and of the stone structures found there; his article was published in 1844. Baer described a stone labyrinth and heaps of stone and discussed their analogues. This was the first work on labyrinths in Russia and the questions were raised as to the purpose and time of the construction of the mazes. Baer was of the opinion that they were constructed by the Finnish or Russian population during the medieval period (Baer 2014). There is also a schematic representation of the labyrinth on Y. Virgin drawn by the local teacher Emil Olenius in 1919 (Uino 2003).

The stone structures were examined in 2019. They are represented by two labyrinths (Fig. 9-10), stone circles and heaps of stone. The structures are located on the upper terrace ca 8 m above sea level. The mazes are about 7,7 m in diameter; their walls are constructed of a single course of stones. The state of preservation of these labyrinths is fairly good but the borders of some of the lines are rather blurred. The distance between the labyrinths is about 50 m. The circles are similar in their dimensions to the labyrinths, and they have 1–2 external walls constructed of one course of stones; in their centres there are small heaps of stone. The stone heaps include two large ones about 1,5 m high and a group of smaller heaps up to 0,5 m in height.

Labyrinths are fairly widely distributed throughout Northern Europe. Usually they are close to the sea, being situated on islands and peninsulas near the shoreline. A small number of them are known in Sweden and Finland, as well as in Russia (particularly on islands in the White Sea). The closest analogues are found among the mazes on the southern coast of Finland: in total, about 200 labyrinths are known to exist in Finland (Uino 2003). Their easternmost examples, according to old Finnish data, are situated on islands of the northern littoral of the Gulf of Finland in the Vyborg region (Bel'skiy 2006).

Conclusion

In the course of the expedition, explorations were conducted on the archipelago of the Outer Islands in the Gulf of Finland, viz. the islands of Moshchny, Maly, Seskar and Yuzhny Virgin. Of most significance are the studies on Moshchny Island, where six sites of archaeological heritage were noted.

The earliest period of the occupation of the islands seems to be that of the Bronze Age. On Moshchny Island we attribute the Suisaari stone cairn to this period. Perhaps at this time the islands were not yet permanently inhabited.

The next period, the early Iron Age, is represented by the cemetery of Suursuonmäki and ceramics of the settlement of Lavansaari (according to the AMS date of the soot on the ceramics). The size and final dating of the settlement of Lavansaari, where the stroked pottery was found, is difficult to determine on the basis of small-scale excavations; it may have been a short-lived settlement. The cemetery of Suursuonmäki contains at least 25 burial mounds and is located in the forest quite far from the seashore. Its size and topography indicate that it belonged to the permanent inhabitants of the island.

The medieval period of the occupation of the archipelago has not been archaeologically investigated, although the first information of written sources about the residents of the islands dates back to the late 14th century, while beginning with the 16th century there are detailed data on the population size on the islands (Korhonen 1996). This probably reflects the research situation and it can be assumed that the data will be identified in further fieldwork. The period of the Middle Ages and the Modern Period comprise the stone labyrinths on the island of Yuzhny Virgin and the cairns on Moshchny Island, as well as the foundation of the church on Moshchny Island.

In terms of their situation, the Outer Islands is equally distant from the northern and southern coasts of the Gulf of Finland. Nevertheless, the sites of different periods on the islands are culturally closer to the northern littoral (Vyborg region of Leningrad oblast and southeastern Finland). The Suisaari burial cairn of the Bronze Age is related to the cultural group of the Bronze Age cairns (*hiidenkiuas*) that are widespread throughout the northern coast of the Gulf of Finland. On the southern coast, in Estonia, a slightly different funerary tradition was practised in the Bronze Age – here stone-cist graves were made (Lang 2007: 147–164; 2018: 162–167). In the early Iron Age, the construction tradition of stone cairns represented on Moshchny Island finds its parallels in the cemeteries of Finland. During the same period on the southern coast of the Gulf of Finland, in Estonia and on the Izhora Plateau, the tradition of making tarand graves was widespread (Lang 2007: 170-219; 2018; Yushkova 2016). The closest parallels to the medieval stone mazes on Y. Virgin are also represented by the stone labyrinths of the northern littoral of the Gulf of Finland; about 200 of them are known in Finland, whereas only a few labyrinths have been reported from Estonia.

The investigations of 2019 demonstrated the importance of the continuation of the archaeological investigation of the islands. When summarising the available data, it must be noted that, at present, no archaeological traces of the permanent medieval settlement of the islands have been found, although its presence is known from written sources. In addition, the question of the time period of the earliest occupation of the islands has also not been solved. In the meantime, sites belonging to even earlier stages of the settlement of the Gulf of Finland region may yet be revealed.

Acknowledgements

I thank A. V. Gilevich for translating the article.

This work was carried out within the framework of the Programme of Fundamental Scientific Research of the State Academies of Sciences, State Assignment No. 018420190007.

References

Asplund, H. 2008. *Kymittæ. Sites, centrality and long-term settlement change in the Kemiönsaari region in SW Finland*. Annales Universitatis Turkuensis. B 312.

von Baer 2014 = К. Э. фон Бэр 2014. Лабиринтоподобные каменные выкладки на Русском Севере. *Сейды, лабиринты, древние камни Арктики*: 181-190. СПб: Гйоль.

Bel'skiy 2006 = Бельский, С.В. 2006. Каменные лабиринты побережья Выборгского залива Балтийского моря (к постановке проблемы). *Первобытная и средневековая история и культура европейского Севера: проблемы изучения и научной реконструкции*: 122–127. Соловки.

Carpelan, C. & Uino, P. 2003. Between Estonia and Finland – a Reassessment of old Finds from Tytärsaari Island. *Arheoloogiga Läänemeremaades*: 75-93.

Edgren, T. 1993. Lavansaaren Suursuonmäen röykkiöhaudat. *Suomen Museo* 1992: 5–20.

Korhonen, M. 1996. Suomenlahden ulkosaarten historiaa keskiajalta 1800-luvun alkupuoliskolle. In *Suomen-*

lahden ulkosaaret – Lavansaari, Seiskari, Suursaari, Tytärsaari: 69-116. Helsinki.
Lang, V. 2007. *The Bronze and Early Iron Ages in Estonia*. Tartu.
Lang, V. 2018. *Läänemeresoome tulemised*. Tartu.
Lavento, M. 2015. Pronssi- ja varhaismetallikausi. In *Haggrén G., Halinen P., Lavento M., Raninen S., Wessman A. Muinaisuutemme jäljet. Suomen esi- ja varhaishistoria kivikaudelta keskiajalle*: 123–212. Helsinki.
Meinander, C. F. 1954. *Die Bronzezeit in Finnland*. Suomen Muinaismuistoyhdistyksen Aikakauskirja.
Miettinen, T. 1996. *Suomenlahden ulkosaarten esihistoria. In Suomenlahden ulkosaaret – Lavansaari, Seiskari, Suursaari, Tytärsaari*: 52-67. Helsinki.
Miettinen 1997 = Миеттинен, Т. 1997. Изменение картины железного века на северном побережье восточной части Финского залива. *Славяне и финно-угры. Археология, история, культура. Доклады российско-финляндского симпозиума по вопросам археологии*: 62-70. Санкт-Петербург.
Miettinen, T. 2012. Pronssikausi ja varhaismetallikausi 1900 eKr. – 300 jKr. In *Kymenlaakson historia*, 1: 53–63. Helsinki.
Saksa 2014 = Сакса, А.И. 2014. Острова восточной части Финского залива в контексте археологических памятников Эстонии, Финляндии и Карельского перешейка. *Бюллетень ИИМК РАН*, №4: 153-158. Санкт-Петербург.
Tuovinen, T. 2002. *The Burial Cairns and the Landscape in the Archipelago of Åboland, SW Finland, in the Bronze Age and the Iron Age*. Acta Universitatis Ouluensis B Humaniora 46.
Uino, P. 1997. *Ancient Karelia. Archaeological studies – Muinais-Karjala. Arkeologisia tutkimuksia*. Suomen Muinaismuistoyhdistyksen Aikakauskirja 104.
Uino, P. 2003. Itäisen Suomenlahden jatulintarhat. *Karjalan synty. Viipurin läänin historia I*: 485.
Yushkova, M. A. 2016. New group of sites of the 1st to 7th centuries AD in the south-west of Leningrad Oblast. In P. Uino & K. Nordqvist (eds.): *New Sites, New Methods. Proceedings of the Finnish-Russian Archaeological Symposium Helsinki, 19–21 November, 2014*. Iskos 21: 143-159.

Notes

1 The Vneshniye Ostrova (Outer Islands) archipelago also includes the islands of Hogland (Suursaari), Bolshoy Tyuters (Tytärsaari) and Maly Tyuters (Pien-Tytärsaari).

2 The results of these excavations are discussed in A. Gorodilov's paper in the present collection.

3 1264-1282 AD (738±25 BP SPb-3063), 1261-1281 AD (745±25 BP SPb-3064) and 1261-1281 AD (745±25 BP SPb-3086).

Some Results of the Multidisciplinary Approach to the Study of Archaeological Sites of Staraya Ladoga

Natalia V. Grigoreva

Natalia V. Grigoreva, Institute for the History of Material Culture, Russian Academy of Sciences, Dvortsovaya emb., 18, 191186 Saint-Petersburg, Russia, e-mail: mak-kon4@yandex.ru

Abstract

In recent years, a multidisciplinary approach has been applied to the results of archaeological excavations in Staraya Ladoga. This article provides a short list of works that have been carried out in the investigations of the occupation layers and findings. The article also provides references to publications based on the results of interdisciplinary research.

In memory of Anatoliy Nikolaevich Kirpichnikov

1 Introduction

Using multidisciplinary methods of investigation is an essential part of modern archaeological research. Not only have the study of occupation layer of archaeological sites and the research of findings found there advanced to a new level, but so has the very important study of the structure of archaeological finds, as well as the process of their deposition. This approach to archaeological material and contexts allows us to answer questions that would be difficult to resolve using only archaeological methods. A multidisciplinary approach makes it possible to construct a detailed multidimensional model of the everyday life of ancient societies, attest changes in environment and landscape, and evaluate the degree of mutual influence of different communities.

Staraya Ladoga, or Ladoga as it was called before 1703, is located 128 km to the east of St. Petersburg, on the left bank of the Volkhov River and 12 km from the southern shore of Lake Ladoga. In the early medieval period Ladoga was one of the most important towns in the border region between northern Russia and the lands to the west.

The focal area of settlement was situated to the south of the cape by the Ladozhka River, on the lower part of the plateau, hidden from the waters of the Volkhov by a coastal dune. This site was investigated during the excavations of E. A. Ryabinin in 1975. The dendrochronology method was used to date the floorings and details of wooden structures from these excavations, located on virgin soil. According to the dendrochronological data, the first buildings appeared here no later than 753 AD (Chernykh 1996: 105-112).

Scandinavian, Slavonic and Finnic findings were found in the occupation layers of the settlement (Ryabinin 1985: 27-75; Kirpichnikov 2014: 215-230).

According to the Hypatian Codex, the legendary Varangian leader Rurik arrived at Ladoga in 862 AD (The Hypatian Codex 1923: 16), and the town was used as a military outpost. This settlement remained attractive for Scandinavian traders, craftsmen and farmers throughout the Viking Age. An

Figure 1. Zemlyanoye gorodishche. View from the south. Photo by S. Vasiljev 2021.

important mention of Ladoga is dated 1019 AD in the chronicles, when Princess Ingegerd Olofsdotter of Sweden married the Grand Prince of Kiev, Yaroslav Mudryy (Yaroslav the Wise). As part of their marriage agreement, Yaroslav ceded Ladoga to his wife, who appointed her father's cousin, the Swedish earl Ragnvald Ulfsson, to rule the town (Melnikova 2013: 533). There was a separate jarlsriki of Ladoga. From the beginning of the 12th century onwards, Ladoga came under the cultural and political influence of Novgorod.

According to Anatoliy N. Kirpichnikov, the first stone fortress was built here at the end of the 9th century (Kirpichnikov 1984: 23-34). However, the chronicle mentions the building of a second stone fortress in 1114 (The Hypatian Codex 1923: 273). During the later periods the fortifications underwent numerous reconstructions. The last of these was conducted in the middle of the 17th century.

The settlement area expanded considerably at the beginning of the 9th century. The houses were built on the right bank of the Ladozhka River. By the end of the 16th century, a part of the settlement was surrounded by earthen walls and merged with the fortress (Selin 2010: 334-344). For this reason, this place started to be called Zemlyanoye Gorodishche (Earthen Hillfort).

The stone fortress and Zemlyanoye Gorodishche continue to be the centre of Ladoga up to the present day. Archaeological excavations are now undertaken in the stone fortress and at Zemlyanoye Gorodishche and other areas of the medieval settlement (Fig. 1). The Staroladozhskaya Archaeological Expedition of the Department of Slavic-Finnish Archeology of the Institute of the History for Material Culture of the Russian Academy of Sciences is working in Ladoga every year. From 1972 to 2013, the head of the expedition was Anatoly N. Kirpichnikov (Fig. 2). All the works that will be discussed

Figure 2. The remains of a wooden building in the cultural layer of the middle of the 9th century. Anatoly Kirpichnikov photographs the hearth. Excavations in 2009. Photo: S. Beletckiy.

in this text were carried out in the southern part of Zemlyanoye Gorodishche during the excavations of 1999-2013.

For many years, the studies of the Ladoga settlement and of the archaeological finds discovered here were comprised of systematic cooperation with specialists of various branches of natural science.

2 The investigation of soil formation of the Neolithic and early medieval period

The study of buried soil (Fig. 3) under the earliest occupation layer of the Ladoga settlement and the features of the early landscape has been carried out together with researchers of the Institute of Geography of the Russian Academy of Sciences. Based on the study of selected samples and the literature data, the several stages of soil formation of Ladoga have been identified, from the Neolithic through the Ladoga transgression and accumulation of sapropel (maximum of transgression 4000 calBP), to the appearance of the Ladoga settlement in the middle of the 8th century AD (Aleksandrovskiy & Aleksandrovskaya 2018: 168-180; Aleksandrovskiy et al. 2010; Aleksandrovskiy et al. 2017; Aleksandrovskiy et al. 2018; Aleksandrovskiy et al. 2019: 21-24). The excavations at Zemlyanoye Gorodishche and the open settlement area revealed traces of farming in the upper part of the soil. This discovery sends us back to the question of the period when the settlement existed. It also opens a discussion of its possible cultural affiliations (Lapshin 2019). Over the course of the study, small pieces of coal from the upper part of the buried soil layer associated with agricultural cultivation were dated from excavations located in the southern and southeastern part of Zemlyanoye Gorodishche. Judging from the results, economic activity and intensive excavations of the site could have been carried out from the beginning of the 5th century AD to the middle of the 8th (Aleksandrovskiy 2012: 60. Tabl. 1).

3 The investigation of archaeobotany

The archaeobotanical studies of the cultural layer of the Ladoga settlement have been traditional ones. At the end of the last century, the Staraya Ladoga archaeological expedition worked in close cooperation with Finnish researchers. The work was carried out in the northwestern part of Zemlyanoye Gorodishche (Aalto & Heinäjoki 1997). Samples were taken from the lower part of the cultural layer from cultural horizons of the 9-10th centuries. A number of industrial crops (common hops (Humulus lupulus), common millet (Panicum miliaceum L.), hemp (Cannabis sativa L.)), grain crops (barley (Hordeum vulgare L.) and emmer (Triticum dicoccum, Triticum aestivum compactum)), as well as individual grains of rye (Secal cereale) and oats (Avena sativa), were discovered from the samples. These excavations made it possible to get an idea of the natural environment of Ladoga in the 9-10th centuries. Archaeobotanical work in recent years has been carried out in the southeastern part of Zemlyanoye Gorodishche and focused primarily on the study of the buried soil. Together with experts of archaeobotanical studies of the N. I. Vavilov All-Russian Institute of Plant Genetic Resources (VIR), 15 samples were obtained from the buried soil and the lower part of the occupation layer, from which almost 19000 fruits and seeds (and their fragments) of cultivated and wild plants were isolated using the flotation method (charred plant remains gathered from wet-sieved soil samples). Samples for

Figure 3. Excavation of the buried soil layer. Excavations in 2010. Photo: S. Beletckiy

research were taken in 2010 and 2013 from the bottom part of the cultural layer and buried soil. The results of the carpological analysis revealed complexes of weed-ruderal, forest-edge, meadow (including marsh-meadow and coastal meadow-marsh) and cultivated plants. Among the macrostates of wild plants (wild plant macrofossils), the seeds of 44 species were identified; some of the remains (2636 pcs) could not be identified due to the insufficient number of preserved morphological features defined only up to the genus or family. Seeds of cultivated plants (1628 pcs) were found in all samples and were assigned to nine taxa: cultivated barley (Hordeum vulgare L.), bread wheat (Triticum aestivum L.), emmer (Triticum dicoccon Schrank), oats (Avena sp.), common millet (Panicum miliaceum L.), hemp (Cannabis sativa L.), peas (Pisum sativum s. l.), flax (Linum usitatissimum L.) and common poppy (Papaver somniferum L.) (Chukhina et al. 2014; Chukhina et al. 2015). Additionally, the analysis of the occupation layer and buried soil which attested to the existence of phytoliths inside them was carried out together with the Department of Botany of Altai State University. Studies have identified several palynocomplexes that characterise vegetation for each of the periods of formation of soil deposits (Speranskaya et al. 2017). The stratigraphic distribution and composition of phytoliths of the occupation layers and virgin soil confirm the previously expressed concepts of the development of landscape and natural conditions in the Lower Volkhov Region, which are closely related to fluctuations in the water level of the Volkhov River. Phytoliths of cultivated cereals, thin-toothed sticks with long, often branching (dendric) teeth that are formed in the spikelet glums of cultivated cereals, were not found in the samples of buried soil. This may mean that, despite the traces of agricultural tillage discovered by soil scientists, no cereals were cultivated at the site of the future settlement. Excavating the soil and applying fertilisers is also common while cultivating garden plants (Platonova et al. 2018).

4 The investigation of osteology and archaeozoology

The study of animal bone remains was carried out by specialists in archaeozoology from the Institute of Plant and Animal Ecology of the Ural Branch of the Russian Academy of Sciences. They investigated the species composition of animals whose bones, ranging from the Viking Age to the Modern Age, were found in excavations in 2006-2013. As a result, it becomes possible to find out the ratio of domestic and wild species, herd structure, age, etc. Interesting results were obtained from materials dating back to the Viking Age. During the 2009 excavations, when the layers of the second half of the 9th century were examined, a significant osteological collection was assembled, including the bones of mammals, birds and fish (Bobkovskaya 2010). The overwhelming majority of mammalian bone remains belong to domestic animals. Most of them belonged to pigs, followed by cattle (cows). These two types were used to produce meat (pork, beef) and dairy products. It is possible that, when breeding cattle, dairy farming was predominant. Small cattle (sheep, goats) occupied the third place in the herd in terms of numbers, and, judging by the nature of bone fragmentation, their purpose was complex, e.g. wool and milk. Perhaps the sheep were used as a sacrificial animal. Fragments of horse bones were found in small amounts. The inhabitants' diet did not include horse meat; horses were kept for household purposes. There were very few cat and dog bones among the finds.

Though commercial mammal species were diverse in composition (ungulates, carnivores, rodents), they were very few among the finds. Beavers and hares could be eaten; the other species (beaver, hare, fox, elk, red deer, wolf, lynx, bear, marten, otter, squirrel) were hunted for their skins, and the Ladoga seal (Pusa hispida ladogensis) was most likely hunted for its fat (blubber). Hunting clearly played a supporting role, but the actual catch volume is likely to be underestimated, since the vast majority of animals caught during hunting were butchered on site, thus their bone remains were not found on the territory of the settlement. Fishing played an important role in the town's economy. Several works by researchers from the Institute of Ecology and Subsoil Use (Kazan, Republic of Tatarstan) are devoted to the study of ichthyofauna from excavations of the 9th-10th-century occupation layer at Zemlyanoy Gorodishche. 23 fish species were identified, among which the most widely represented are the pike perch (Sander lucioperca), the sturgeon (Acipenser oxyrinchus), the bream (Abramis brama), the pike (Esox Lucius) and the catfish (Silurus glanis) (Galimova et al. 2015). Genetic studies of these species were also carried out. Part of the collection of finds of early medieval bird bones was also analysed at the Institute.

Bird bones from the excavations, which were carried out in the area of the stone fortress, were processed by researchers from the Zoological Institute of the Russian Academy of Sciences and the Department of Genetics and Biotechnology of St. Petersburg State University. 158 bird bone samples from excavations in the stone fortress performed during 2014-2015 were analysed. We determined that 108 bone remains belonged to 37 individuals of 16 species. In terms of the number of bones and the number of individuals in the layers of different times, the remains of waterbirds and chickens prevail (Galkina et al. 2016).

Domestic chickens (Gallus gallus domesticus) were abundantly represented: more than 40% of all recognisable bones belonged to nine individuals of varying ages. We also found bones of wild galliform species that were probably eaten, such as black grouse (Lyrurus tetrix) and capercaillie (Tetrao urogallus).

Waterbirds were the most diverse group in terms of the number of species identified. We found bones (39%) belonging to the common swan (Cygnus cygnus), graylag goose (Anser anser), mallard (Anas platyrhynchos), common teal (Anas crecca), northern shoveler (Anas clypeata) and red-breasted merganser (Mergus serrator). It is worth noting that mallards were commonly present in excavated bone samples (20% of all bones) and one of the three graylag goose individuals found was a young bird, no older than 1 month. Still, it is hard to determine which were caught in the wild and which were domesticated.

There are bone remains of three individuals of the common eider (Somateria mollissima) among the material dated to the 10th-12th centuries. This species breeds in the Arctic and some northern temperate regions, and its nests can be found on the White Sea coasts, as well as on islands of the Baltic Sea and in the very west of the Leningrad region. Rare migrations of the common eider of the western population to St. Petersburg were recorded in May, and on non-freezing areas of the Neva in winter. The eider is an extremely rare species on Lake Ladoga, and it is not found in small inland waters. Birds of the northern population were registered only a few times in autumn in the north and west of Lake Ladoga. Perhaps in the 10th-12th centuries the habitat of this species was more favourable and it nested on the northwestern rocky islands of Lake Ladoga, and roamed the Ladoga area during the non-breeding period.

Exclusively in mixed layers, we found bones of birds of prey. The buzzard (Buteo buteo) is a medium-sized bird that could be hunted for meat. Sparrowhawks (Accipiter nisus) were probably buried entirely. They could be killed to protect domestic chickens or used as a hunting bird, and then buried after death. The bones of an eagle-owl (Bubo bubo) have also been identified.

In addition, bone remains of a jackdaw (Corvus monedula), a common raven (Corvus corax, Passeriformes) and a common crane (Grus grus, Gruiformes) were identified, which probably served as hunting trophies.

5 The investigation of anthropology

Recently, additional research was undertaken on skeletal remains excavated from a Christian burial site dating back to the 11th-

12th centuries. The cemetery is situated in the central part of Zemlyanoye Gorodishche near the church of Saint Clement (Ravdonikas 1949: 12; Platonova & Sankina 2018: 101-134). The special anthropological features of the people buried show their possible Scandinavian origin. The cemetery was discovered in 1938–1940 during an excavation by V. Ravdonikas and P. Grozdilov. Physical anthropologist Alexandr N. Yuzefovich, who participated in the excavations, wrote in 1941 that the cranial series discovered among the remains is different from those of Slavonic stock. In 1969 it was noted that the trait combination setting this group apart from the Eastern Slavonic one was paralleled by that observed in Iron Age groups of Scandinavia (Platonova & Sankina 2018: 107). A new study of Staraya Ladoga skeletons (both cranial and postcranial bones), undertaken by S. Sankina in the 1990s, demonstrated chronological and typological differences in another part of this cemetery. In this place (apparently the earliest one), there were young males (mostly below the age of thirty or juvenile) and children, including infants. Using the radiocarbon method of analysis, the bone remains of several burials were dated. As a result, several dates from the 10th-12th centuries were obtained. A strontium isotope analysis from 15 burials in this part of the cemetery was conducted in the Laboratory for Archaeological Chemistry at the University of Wisconsin-Madison. The isotope analysis of the teeth of buried persons has shown some distinction from local fauna samples and identified at least five non-local individuals with isotopic signatures that match the Mälaren region of east-central Sweden (Price et al. 2019).

In the last decade, major archaeological work took place in the area of the stone fortress, which housed a parish cemetery near St. George's Chuch from the 18th to the early 20th centuries. The anthropological findings were investigated with anthropologists from the Peter the Great Museum of Anthropology and Ethnography (the Kunstkamera), Russian Academy of Sciences (Moiseev et al. 2016). The remains of more than 100 persons from the 18th-20th centuries were discovered. The finds are similar in their craniometric characteristics to those originating from the excavations of the northern part of the cemetery, dating to the 17th-18th centuries, on Zemlyanoye Gorodishche.

6 Geoarchaeological investigations

New surprising results were obtained during the study of the courtyard of the stone fortress in 2017 using high-precision gravimetry. The study was conducted with A.P. Karpinskiy, with support from the Russian Geological Research Institute (VSEGEI), in 2017. It allowed us to conclude that the coastal cape was initially much smaller than today (Grigor'yeva et al. 2018: 195). A stone wall with wooden pillars of the most ancient fortress (in the 9th century) stood along the border of the cape (Grigor'eva 2020). During the 11th and up to the beginning of the 12th century, the eastern side of the cape was augmented with earth brought from outside, which significantly increased the area of the medieval fortress. Without gravimetrical investigations, the complete picture of the changes would remain unclear.

7 The investigation of the chemical composition of finds

Every year, during excavations at the settlement, a large number of finds of various materials are discovered (glass, metal, iron, ceramics, stone, wood, leather). Different categories of archaeological finds made from glass and metal alloys were set up, using the methods of chemical, X-ray, structural and microprobe analysis. Due to the use of the spectral analysis method (Laboratory of spectral analysis of the IIMC RAS), it was found that a glass workshop for the production of glass beads from imported raw materials functioned in Ladoga from the second half of the 10th to the beginning of the 11th century. This conclusion is confirmed by the chemical composition of the glass belonging predominantly to the soda-lime-silica type, which is specific for Byzantine glass composition. Cobalt was the principal colourant, but some yellow opaque lead-containing glass was also revealed. On the whole, the Staraya Ladoga glassworking is typical for Northern Europe at this time (Grigoreva & Lesman 2012; Grigoreva & Egor'kov 2020). In 2012, a spall of sintered quartz sand, covered at one side with a thin layer (less than 3 mm) of transparent colourless glass, was unearthed in excavations at the southeastern part of Zemlyanoye Gorodishche. This find allows us to suggest that the first attempts of glass melting from the local source materials were made in the Ladoga workshop (Grigorieva & Kulkova 2020).

Ladoga was a centre of redistribution of metal raw materials; a jewelry workshop was located and constantly functioned on the territory of the settlement. Tools (pincers, crucibles), metal ingots with traces of use, and semi-finished and defective products are widely represented among the archaeological finds. There are also fragments of melted metal that for some reason were not used in production. One such copper ingot was selected for lead isotope analysis. According to the results of the isotopic analysis of lead, the copper of the found ingot turned out to be close to the composition of lead from deposits in Austria and Germany (Grigoreva et al. 2021). Such finds confirm the high trade status of ancient Ladoga, and the unclaimed metal shows that there was no shortage of it.

8 Conclusion

Thus, close cooperation with researchers of various profiles allows us to clarify the facts of life in Ladoga, including details that we have not thought about before: when did poultry appear in Ladoga, or to what extent was poultry farming developed in the Northwest? How did human activities influence the formation of the surrounding landscape, and to what extent did the economy of Ladoga depend on spring floods? Did technological advancement allow for the production of glass from local quartz sand? As a result, such multidirectional studies contribute to the creation of a general picture of the world of medieval Ladoga and its inhabitants. Collaboration in close cooperation with specialists of different profiles provides a lot of new information that allows us to correct and expand on the existing notions. As part of the work of the Staraya Ladoga archaeological expedition, the plan is to continue and expand the existing cooperation with specialists in the natural sciences.

References

Aalto & Heinäjoki 1997 = Аалто, М & Хейнайоки, Х. 1997. Растительность и окружающая среда Старой Ладоги в эпоху викингов. In А. Н. Кирпичников & Е. Н. Носов (eds.) *Древности Поволховья*: 31-41.

Alexandrovskiy 2012 = Александровский А. Л. 2012. Радиоуглеродный возраст пахотного горизонта под культурным слоем Земляного городища Старой Ладоги. *Ладога и Ладожская земля в эпоху средневековья* 3: 59-61.

Alexandrovskiy & Alexandrovskaya 2018 = Александровский А. Л. & Александровская Е. И. 2018. In Н. И. Платонова & В. А. Лапшин (eds.) *Новое в археологии Старой Ладоги. Материалы и исследования*: 168-180. Труды ИИМК РАН. Вып. LIII.

Alexandrovskiy et al. 2010 = Александровский А. Л., Кренке Н. А. & Нефёдов В. С. 2010. Исследования почв и отложений под культурным слоем Земляного городища Старой Ладоги In: Носов Е. А. & Белецкий С. В. (eds.) *Краеугольный камень. Археология, история, искусство, культура России и сопредельных стран*: 43-61. Том I.

Alexandrovskiy et al. 2017 = Александровский А. Л., Лапшин В. А., Григорьева Н. В., Александровская Е. И. & Скрипкин В. В. 2017. Старая Ладога: палеопочвы и природная среда в неолите и раннем средневековье. In Н. О. Ковалева, С. К. Костовска, А. С. Некрич & О. А. Салимгареева (eds.) *Человек и природа*: 10-14. Материалы XXVII Международной междисциплинарной конференции 'Проблемы социоестественных исследований' и Международной междисциплинарной молодежной школы 'Стратегии экологической безопасности'. http://www.igras.ru/sites/default/files/%D0%9C%D0%B0%D0%BA%D0%B5%D1%82%20%D0%A1%D0%95%D0%98%20 2017%20%D0%BE%D0%BA%D0%BE%D0%BD%D1%87%D0%B0%D1%82%D0%B5%D0%BB%D1%8C%D0%BD%D1%8B%D0%B9.pdf Read 10.09.2020.

Alexandrovskiy et al. 2018 = Александровский А. Л., Лапшин В. А., Григорьева Н. В., Александровская Е. И., Скрипкин В. В. 2018. Почвообразование на территории Старой Ладоги в неолите и раннем средневековье. *Записки ИИМК РАН* 18: 99-104.

Alexandrovskiy et al. 2019 = Александровский А. Л., Ершова Е. Г., Лапшин В. А., Григорьева Н. В., Долгих А. В. & Кирпичников А. Н. 2019. Палеоэкология голоцена и этапы заселения Старой Ладоги в неолите – раннем средневековье. In Савинецкий А. Б. (ed.) *Динамика экосистем в голоцене*: 21-24. Материалы V всероссийской конференции с международным участием 'Динамика экосистем в голоцене' (к 100-летию Л. Г. Динесмана).

Bobkovskaya 2010 – Report about the excavations of the Staroladogskaya archaeological expedition of IIMK RAS in the Staraya Ladoga , Volkhovsky district , Leningrad region in 2009. Volume 1. The text of the report and the appendix. Appendix 5. Scientific Archive of Institute for the history of material culture. Manuscript department. F. 35, in. 1/2009, f. 100.

Chernykh 1996 = Черных Н. Б. 1996.*Дендрохронология и археология*. Москва: Институт археологии РАН.

Chukhina et al. 2014 = Чухина И. Г, Радыш Д. В. & Григорьева Н. В. 2014. Предварительные результаты изучения растительных макроостатков по материалам раскопок 2010 года в Старой Ладоге. In *Ладога в контексте истории и археологии Северной Евразии*: 244-251. Материалы XVII чтений памяти Анны Мачинской и XVIII Чтений памяти Анны и Дмитрия Алексеевича Мачинских.

Chukhina et al. 2015 = Чухина И. Г., Радыш Д. В. & Григорьева Н. В. 2015. Comparison of the Staraya Ladoga Earthern Gorodishche and Gorodishche Kamno Carpological Research Results. (In Russian). *Археология и история Пскова и Псковской земли* 30 (60):223-229.

Galimova, D. N., Askeyev, I. V. Askeyev, O. V., Popović, D. & Panagiotopoulou, H. 2015. A study of the fish bones from the medieval town - Staraya Ladoga. Fishing through time. Archaeoichthyology, Biodiversity, Ecology and Human Impact on Aquatic Environments. 18th Fish Remains Working Group International Council for Archaeozoology. 28th September - 3rd October 2015 Lisbon: 63-64.

Galkina et al. 2016 = Галкина С. А., Демин А. Г., Пантелеев А. В., Дружкова А. С., Трифонов В. А. & Григорьева Н. В. 2016. Перспективы и методика анализа ДНК ископаемых костей домашней курицы, обнаруженных при археологических раскопках. In В. Т. Мусбахова (ed.) *Ладога и проблемы древней и средневековой истории северной Евразии*: 118-124. Сборник статей по материалам XIX и XX чтений памяти Анны и Дмитрия Алексеевича Мачинских.

Grigor'eva 2020 = Григорьева, Н. В. 2020. Новые данные о ранней Ладожской крепости. Хронология и стратиграфия (результаты раскопок около Раскатной башни). *Археологические вести*. 30: 134-153. In print.

Grigoreva & Egor'kov 2020 = Григорьева Н. В. & Егорьков А. Н. 2020. Химический состав изделий и отходов стеклоделательной мастерской Старой Ладоги конца X - XI века. *Yu. L. Shchapova International Memorial Conference*: 45. http://www.hist.msu.ru/upload/iblock/d83/%D0%A2%D0%B5%D0%B7%D0%B8%D1%81%D1%8B%20%D0%BA%D0%BE%D0%BD%D1%84%D0%B5%D1%80%D0%B5%D0%BD%D1%86%D0%B8%D0%B8.pdf Read 10.09.2020.

Grigorieva, N. M. & Kulkova, M. A. 2020. New Data in Medieval Ladoga Glass-Making. *Book of Abstracts international scientific conference Glass along the Roads of Eurasia in Ancient Times and Medieval period*: 33-34.

Grigoreva & Lesman 2012 = Григорьева Н. В. & Лесман Ю. М. 2012. Новые материалы о стеклоделии в средневековой Ладоге. *Ладога и Ладожская земля* 3: 62-77.

Grigoreva et al. 2021 = Grigoryeva, N. V., Krupenik, V. A. & Krupenik, Z. V. 2021. Copper from banks of the Volkhov River. (In Russian) *Археологические вести*. 31:277-285.

Grigor'yeva et al. 2018 = Григорьева Н. В., Кошевой В. В. & Медведев О. Ю. 2018. Археолого-геофизические исследования на территории Ладожской крепости (опыт применения метода высокоточной гравиметрии для получения данных о рельефе местности). *Археологические вести*. 24: 191-197.

Kirpichnikov 1984 = Кирпичников А. Н. 1984. *Каменные крепости Новгородской земли*. Ленинград: Наука.

Kirpichnikov, A. N. 2014. Early Ladoga during the Viking Age in the Light of the International Cultural Transfers. In A. E. Musin and B. Pierr (eds.) *Vers l'Orient et vers l'Occident*: 215-230.

Lapshin 2019 = Лапшин В. А. 2019. Ладога до Ладоги. *Записки ИИМК РАН* 20: 112-120.

Melnikova 2013 = Мельникова Е. А. 2013.*Древняя Русь в свете зарубежных источников*. Москва: Университет Дмитрия Пожарского.

Moiseev et al. 2016 = Моисеев, В. Г. & Григорьева Н. В. & Широбоков, И. Г. & Хартанович, В. И. 2016. Краниологические материалы из раскопок у церкви Святого Георгия в Старой Ладоге – *Радловский сборник. Научные исследования и музейные проекты МАЭ РАН в 2015 г.*: 390-399.

Platonova & Sankina 2018 = Платонова Н. И. & Санкина С. Л. 2018. Стратиграфия верхних горизонтов Земляного городища Старой Ладоги и грунтовый могильник у церкви св. Климента (по материалам раскопок 1938-1940 гг.) In Н. И. Платонова & В. А. Лапшин (eds.) *Новое в археологии Старой Ладоги. Материалы и исследования*: 66-134. Труды ИИМК РАН. Вып. LIII. Санкт-Петербург: Невская книжная типография.

Price, D. T., Moiseyev, V. & Grigoreva, N. V. 2019. *Vikings in Russia: Origins of the Medieval Inhabitants of*

Staraya Ladoga. Archaeological and Anthropological Sciences 11: 6093–6109.

Ravdonikas 1949 = Равдоникас В. И. 1949. Старая Ладога. *Советская археология*. XI (1): 5-54.

Ryabinin 1985 = Рябинин Е. А. 1985. Новые открытия в Старой Ладоге. In В. Седов (eds.) *Средневековая Ладога*: 27-75.

Selin 2010 = Селин А. А. 2010. О времени строительства Земляного города в Ладоге. In: Носов Е. А. & Белецкий С. В. (eds.) *Краеугольный камень. Археология, история, искусство, культура России и сопредельных стран*: 334-349. Том. II.

Speranskaya et al. 2017 = Сперанская Н. Ю., Чухина И. Г., Григорьева Н. В., Соломонова М. Ю. 2017. Итоги изучения фитолитов погребенной почвы Земляного городища Старой Ладоги. *Сборник научных трудов Всероссийского археологического съезда V (XXI)*: 984-985.

The Hypatian Codex. PSRL [Complete Collection of Russian Chronicles]. Vol. 2. Issue 1. Third edition. Edited by Schakhmatow A. Petrograd. 1923. http://psrl.csu.ru/toms/Tom_02.shtml Read 10.09.2020.

Here be Reindeer: Geoarchaeological Approaches to the Transspecies Lifeworlds of the Sámi Reindeer Herder Camps on the Tundra

Oula Seitsonen & Natalia Égüez

Oula Seitsonen (Sakarin-Pentin Ilarin Oula), Archaeology, University of Oulu, Oulu, Finland, oula.seitsonen@gmail.com ORCID 0000-0003-3792-0081
Natalia Égüez, Archaeological Micromorphology and Biomarkers – AMBI Lab, Instituto Universitario de Bio-Orgánica Antonio González. Universidad de La Laguna, 38206 Tenerife, Spain. neguezgo@ull.edu.es

Abstract

In this paper we present ongoing geoarchaeological research of Sámi reindeer pastoralist campsites. The discussion is based on three case study sites situated at the Lake Gilbbesjávri region (Finland) in northern Sápmi. All the sites are historical-period reindeer herder camps, likely used as temporary summer encampments by the nomadic herders in the 17th–19th centuries. The sites were prospected with systematic soil sampling and geoarchaeological analyses, which have been previously successfully applied in the study of Sámi habitation sites on both intrasite and intra-dwelling levels. The changes in the soil phosphate, pH and electric conductivity values were analysed and interpreted based on ethnographic analogies. These allow insight into the outwardly invisible taskscapes and spatialities of the herders and their animals at the campsites. The different geoarchaeological factors highlight differing aspects of the use of space at the historical Sámi reindeer pastoralist campsites, such as potential reindeer corralling areas, and provide comparative material for the analyses of prehistoric Sámi hunter-gatherer-herder sites.

1 Introduction

The materialities of soils and their geochemical and geophysical properties can act as an archive and proxy for past human and animal activities. In this paper, we approach the archaeological remains of historical Sámi reindeer pastoralists in northern Finnish Lapland from a geoarchaeological perspective. Nomadic reindeer pastoralism and the connected long-distance mobility developed in northern Fennoscandia over the course of the 14th–19th centuries CE from the previous small-scale draught and decoy reindeer herding practiced in a hunter-gatherer-herder society (e.g. Hansen & Olsen 2014). This change took place at a differing pace and with differing details in the various parts of Sápmi, the homeland of the Sámi people covering the northern reaches of Norway, Sweden, Finland and Russia (Fig. 1).

In this paper we present three geoarchaeological case studies from the Mountain Sámi habitation sites situated in the Arctic tundra environment in northernmost Finnish Sápmi in the Lake Gilbbesjávri area (Fi. Kilpisjärvi; Northern Sámi placenames are used throughout). These aim at assessing the outwardly invisible, transspecies taskscapes of the herders and their animals at the campsites through systematic soil sampling and geoarchaeological analyses. The ana-

Figure 1. Top: Location of the studied sites in Sápmi. Studied sites in red, other Sámi sites in black, and background samples with black crosses. Map: Oula Seitsonen; Bottom: View over the 18th–19th century pastoralist campsite Gahperusvárri itä 2, on the foreground a *bearpmet* hearth. Photo: Oula Seitsonen.

lysed geoarchaeological soil factors, such as phosphate concentrations and electric conductivity, have potential to highlight different aspects of the spatialities of the historical Sámi reindeer pastoralist camps, e.g. the use of space inside the dwellings or potential reindeer corralling areas. In many parts of Sápmi, reindeer pastoralism has remained an important livelihood and cultural practice to this day, such as in the Lake Gilbbesjávri region. However, herding practices have changed considerably over the last century, influenced by the emerging aspects of modernity such as permanent housing, all-terrain vehicles, helicopters and satellite navigation (see Nykänen & Valkeapää 2016; Näkkäläjärvi 2013; Seitsonen 2021a).

Geoarchaeological soil analyses have been successfully applied in the study of the Sámi reindeer herders' camps and sacred *sieidi* sites for decades (e.g. Äikäs 2015; Carpelan & Lavento 1996; Halinen 2009; Nuñez 1977). Soil phosphate analysis is one of the most commonly used geochemical approaches in archaeology globally (e.g. Lavento 2003; Linderholm 2010; Trinks 2014) and has often been applied for the intra-dwelling and intrasite studies of Sámi sites (e.g. Halinen 2016; Jerand et al. 2016; Karlsson 2006). Magnetic susceptibility and loss on ignition values are also commonly used analyses (e.g. Jerand & Linderholm 2019; Tolonen 2013), on top of which Nuñez (1977) has analysed pH values from medieval Sámi winter camps and Tolonen (2013) from the sacrificial Sámi *sieidis*. Recently, Jerand and Linderholm (2019) have developed advanced analytical approaches in phosphate analysis by assessing separately both inorganic and organic phosphate contents and their interrelation. The previous studies have assessed, for instance, the intra-dwelling organisations of Sámi tents (Sán. *lávvu* and *bealjegoahti*) and turf huts (Sán. *darfegoahti*) (e.g. Halinen 2016; 2019; Jerand et al. 2016), and the intrasite patterns, such as potential reindeer corralling and gathering areas at the sites (e.g. Jerand & Linderholm 2019; Karlsson 2006).

Our study is part of an ongoing research on the effects that the reindeer domestication and herding have had on the materialities of soils in northern Fennoscandia, as part of the wider project Domestication in Action which studies the human-reindeer relations from multidisciplinary perspectives. This paper provides geoarchaeological case studies from three seasonal pastoralist campsites dated to the 17th-19th centuries. These likely functioned as transitionary summer camps related to the movement of the reindeer herds and herders from their inland winter pastures to the coastal summer pastures (Fig. 1:1–3). In our soil analyses, we have analysed the phosphates, pH and, as a new variable, the electric conductivity (EC) values from these sites. EC analysis of soils was included in the study as an exploratory factor to assess how it might mirror past activities, such as fires or waste disposal, that can alter the EC properties of the soils (e.g. Trinks 2014). Both the intra-dwelling and intrasite patterning of the studied geoarchaeological factors are used to assess human and animal activities and their spatial configurations. The spatialities are approached from a phenomenological perspective as part of the lived-in, transspecies lifeworlds (Husserl 1970 [1936]: 103; Seamon 1979; von Uexküll 1992 [1934]), taskscapes (Ingold 1993) and meshworks (Ingold, 2011: 63–97) of the herders, their reindeer, and the other beings. Intra-dwelling organisations are compared with the well-known and often-cited ethnographic analogies related to the floor plans of the Sámi dwellings (e.g. Itkonen 1948a: 183–184; Jerand et al. 2016; Solbakk 2007: 74). Intrasite spatialities are interpreted based on the historical waste disposal and reindeer handling patterns at the Sámi sites. In addition to being expressive of the lifeworlds of the historical nomadic pastoralists and their animal companions, our study provides also analogies and new comparative data for assessing the spatialities of the Iron Age and medieval hunter-gatherer-herder Sámi campsites (e.g. Halinen 2009; 2019; Halinen et al. 2013; Jerand & Linderholm 2019; Karlsson 2006).

2 Studied reindeer pastoralist sites

We discuss here three systematically sampled and analysed historical Mountain Sámi reindeer pastoralist encampments: Nuvkkágieddi (Fig. 1:1; Fig. 2A), Gahperusvárri itä 2 (henceforth Gahperusvárri; Fig. 1:2; Fig. 2B) and Čáhkaljohka etelä 2 (henceforth Cáhkaljohka; Fig. 1:3; Fig. 2C) (Table 1). These are all situated in the barren roadless

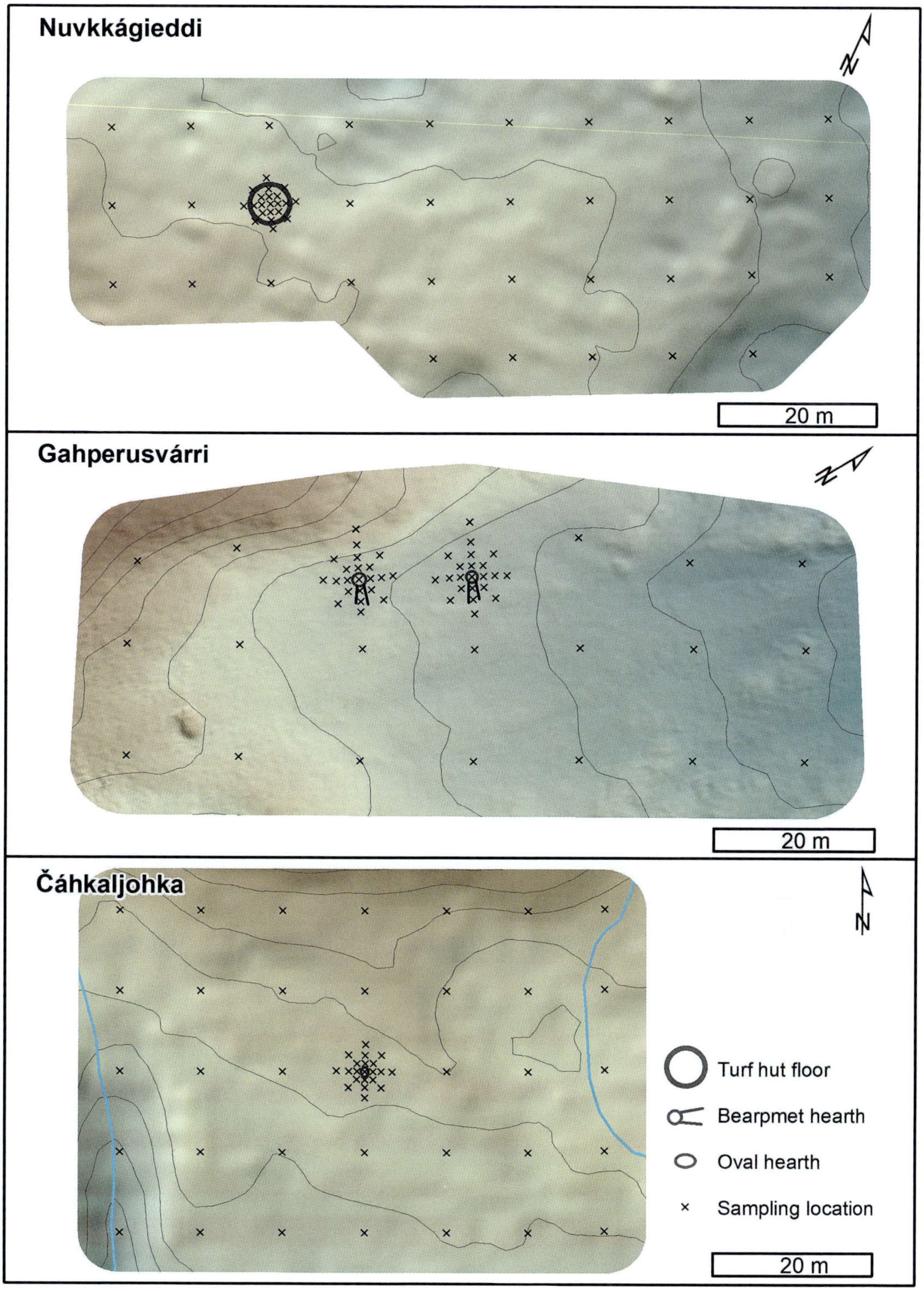

Figure 2. Studied sites, structures and soil sampling locations, elevation contours at 0,5-metre intervals. Maps: Oula Seitsonen.

Site	Type	Environmental context	Meters a.s.l.	Radiocarbon dates	Lab. Code	CalAD (2-sigma)			
						BP	Max	Min	Median
Gahperusvárri	Three hearths	Alpine tundra	760	Oval hearth	D-AMS-036582	204±26	1656	...	1775
Gahperusvárri	Three hearths	Alpine tundra	760	Bearpmet hearth	D-AMS-035325	143±22	1679	1939	1804
Čáhkaljohka	Single hearth	Mountain birch forest	500	Oval hearth	D-AMS-035338	100±22	1690	1926	1841
Nuvkkágieddi	Turf hut foundation	Alpine tundra	603	-	-	-	-	-	-

Radiocarbon dates calibrated with Oxcal v. 4.3.2 (Bronk Ramsey 2017) and IntCal13 atmospheric curve (Reimer et al. 2013)

Table 1. Site characteristics and the radiocarbon dated contexts.

landscape east of Lake Gilbbesjávri (N69°2'/ E20°48') that is still an important reindeer herding country for the local *siidas* (*siida* is a traditional Sámi social and herding unit typically consisting of family and friends) (Fig. 1). Čáhkaljohka is situated in the upper edge of sparse mountain birch forest, whereas the two other sites are in the open Alpine tundra environment (Fig. 1; Table 1).

There are usually few visible material traces on the surface at the mobile Sámi pastoralist campsites, most often only some scattered hearth stones or round hut floors with shallow remains of turf walls (Fig. 1). Then again, vegetation at the sites is often notably different from their surroundings, with more graminoids and less shrub (Fenger-Nielsen et al. 2019; Karlsson 2006: 163; also Itkonen 1948a: 274, for ethnographic examples of the vegetation change). We chose sites with different types of archaeological features connected to reindeer pastoralism to compare their geoarchaeological characteristics (see Karlsson 2006). At Nuvkkágieddi the visible remains consist of a round turf hut foundation, at Čáhkaljohka of an oval hearth, and at Gahperusvárri of two *bearpmet* hearths and one round hearth (with the last one situated outside the area mapped in Fig. 2C) (Table 1). B*earpmetárran* are hearths that typically have two stone 'arms,' *uksabearpmet*, leading from the corners of the hearth to the doorway of the tent, related to the traditional floor plan of the Sámi dwellings. Sometimes logs were used instead of stone lines for the internal division of the dwellings (Itkonen 1948a: 183–184; Ränk 1949), as is still done nowadays at the seasonal herding sites (Seitsonen 2020).

Local transgenerational memories suggest that two of the sites, Nuvkkágieddi and Čáhkaljohka, acted as pastoralist summer camps in the 19th century, and probably already earlier. An oval hearth at Čáhkaljohka was radiocarbon dated to the mid-19th century (Table 1), in agreement with the local oral histories. These sites likely acted as stopover points along the seasonal migration routes between the winter pastures in the Finnish interior and the summer pastures on the Norwegian coast before the national borders were closed from the migrating Sámi after the mid-19th century (e.g. Itkonen 1948b: 225–26; Seitsonen 2021a). After the border closures, many families were forced to move in search of new pasturelands in the southern and eastern reaches of Sápmi (e.g. Heikkinen et al. 2005; Valtonen 2016).

Based on the indigenous histories of one Sámi family, Nuvkkágieddi was used by them before the family was forced to migrate several hundred kilometres southeast in search of new reindeer pastures in another part of Sápmi and adapt to a completely different, forested herding environment in the late 19th century. According to them, they lived at the site in the summers for some time after the borders were closed and the *siidas* were still adapting to the new territories (personal communication with the family representa-

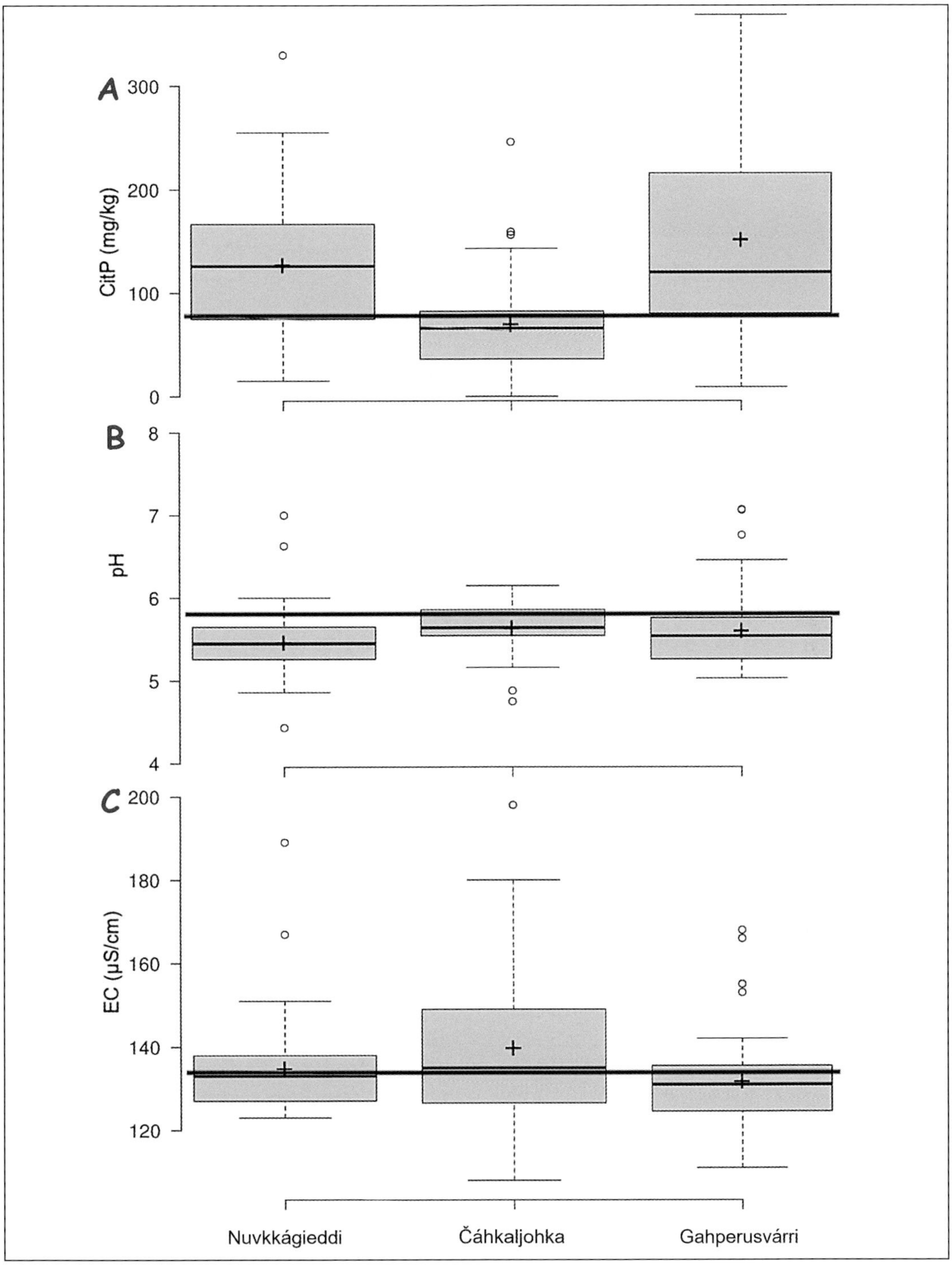

Figure 3. Box-and-whisker plots of the measured geochemical values at each site: A. Phosphates. B. pH. C. Electric conductivity. The boxes illustrate the first quartile, median and third quartile, the whiskers show the 1.5 times interquartile range, and the open dots are outliers. Median values of the background control samples are shown with a thick horizontal line. Figure: Oula Seitsonen.

Phosphate (mg/kg)	Nuvkkágieddi	Čáhkaljohka	Gahperusvárri	Background
Median	126	66	120	76
Mean	126	69	151	77
Min	15	0	9	15
Max	330	246	369	127
pH				
Median	5,5	5,6	5,5	5,8
Mean	5,5	5,6	5,6	5,7
Min	4,4	4,8	5,0	5,1
Max	7,0	6,2	7,1	6,3
EC (µS/cm)				
Median	133	135	131	136
Mean	135	140	132	126
Min	123	108	111	48
Max	189	198	168	142

Table 2. Descriptive statistics of the georchaeological factors measured at the studied sites (Fig. 1:1–3) and at the background samples (Fig. 1) (full results of the analyses are presented in Seitsonen 2021b).

tive). The Gahperusvárri site was tentatively connected by the locals to the former *Pienni siida* that herded their reindeer in this area before they merged with the current *siidas* in the early 20th century (also Itkonen 1948a: 230; 1948b: 159, 166). At Gahperusvárri, a round hearth was radiocarbon dated to the late 18th century CE and a *bearpmet* hearth to the early 19th century CE (Table 1).

3 Materialities of soil: geoarchaeological analyses of the sites

Soils offer an archive of past human and animal activities that can be approached e.g. by studying the chemical and biochemical properties of archaeological sites (Égüez and Makarewicz 2018; Jerand & Linderholm 2019). Soil samples were taken at all three sites with a borer at 10-metre intervals across the whole site and at 1-metre intervals around the hearth structures (Fig. 2). All soil samples were taken from the top part of the enriched B-horizon of the shallow podzol soils. The archaeological deposits at all sites are less than 8 centimetres thick, starting immediately under the turf layer that is only a couple of centimetres thick. The soils at all sites are fine-grained sands underlain by more gravelly material. The phosphate values of the samples were analysed using a citric acid method analogous to that described by Jerand et al. (2016). This allows cautious, relative cross-site comparisons, keeping in mind that the soil properties are highly dependent on various local factors. pH and EC values were measured from sieved samples diluted into distilled water using a Hanna Instruments HI98129 Combo meter (with an HI 73127 pH electrode, and calibrated using the HI7007 and HI7031 solutions; each sample was measured three times and a mean of these measurements was calculated). Nine background control samples were collected from randomly selected localities in the wider landscape (see Fig. 1), providing comparison to the values measured at the archaeological sites. The spatiality of the studied values was visualised using the natural neighbour interpolation of the data points.

The measured geoarchaeological variables from each site and background samples are summarised in Table 2 and illustrated as box-and-whisker plots in Figure 3. Čáhkaljohka stands out from the other two sites, partly due to its differing environmental setting. However, the phosphate values also suggest less intensive use of the site. Most phosphate values at Čáhkaljohka are below the median of the control samples, whereas at the other sites the majority are above it. Compared to the Iron Age and medieval Sámi winter campsites (Halinen 2016; 2019; Jerand et al. 2016), the phosphate values from all our sites are considerably lower. On the other hand, they are reminiscent of the readings from the Utsjoki Kalddasjohka 4 site with a single round Iron Age hearth (Halinen 2008a), the even lower phosphate values measured at the Utsjoki Jeagelveijohka 1 site with several *bearpmet* hearths (Halinen 2008b), and the low values from the pastoralist sites studied by Karlsson (2016: 96–98). These sites with low phosphate values were likely settled seasonally for relatively short durations (see Jerand et al. 2016).

The majority of the pH values from all the archaeological sites are below the background control samples' median value. This is in line with the observations made by Nuñez (1977: 57) of more acidic soils at the Sámi winter camps than in their surrounding natural soils. It is noteworthy that the highest pH readings at all three sites are encountered in and immediately around the hearths, likely related to the alkaline wood ash disposal and perhaps the processing and accumulation of the alkaline bone waste in these areas (Lavento 2003: 52). EC measurements from all sites are close to the background values, with the highest values encountered inside the hearths, and likely linked to both fire keeping and bone waste disposal. Fragments of burnt bones are often located inside the hearths studied in this area, as also elsewhere in Sápmi (see Halinen 2008b; Halinen et al. 2013).

4 Intrasite spatialities

The distribution maps of the measured geoarchaeological readings are presented in Figures 4–6. These illustrate clearly how the dwellings stand out at all sites, with the highest values of all the components originating from inside the hearths. A strong correlation has been noted between the archaeologically deposited bone material and high phosphate values at Sámi sites (Jerand et al. 2016: 365; Jerand & Linderholm 2019: 33). Likewise, the high pH values appear to relate to the deposits that are rich in bone material and ash (Lavento 2003), such as hearths. Thus, the correlations between higher phosphate and pH values at the sites could point to areas where bone waste was handled and dumped, and likely also wood ash.

Areas with both high phosphate and pH values appear e.g. at Nuvkkágieddi about 30 metres in front of the turf hut floor, at Čáhkaljohka northeast of the hearth, and at Gahperusvárri west of the hearths (Fig. 4–6). At both Čáhkaljohka and Gahperusvárri the potential bone handling areas are situated behind the dwellings, as inferred from the setting and structure of the hearths and the geoarchaeological values surrounding them (see below). This is comparable to the observations made at other Iron Age and medieval Sámi sites where garbage dump zones appear typically behind the tents and turf huts (Halinen 2008b; 2019; Hedman & Olsen 2009; Jerand et al. 2016). Also, the ethnographic observations about the waste management in the mobile Sámi pastoralist camps in the early 20th century suggest that both garbage dumps and toilet areas were situated behind the dwellings, often on the right side as seen from the entrance, and the men moved there around the right side and the women around the left side of the dwelling (Itkonen 1948a: 185). Then again, at Nuvkkágieddi the areas with higher phosphate and pH values are about 30 metres in front of the turf hut floor. Some differing waste management pattern might have been practiced at this site, with a more permanent turf-covered hut, than at the hearth sites where mobile tents were used (see Halinen 2019; Carpelan & Lavento 1996; for studies of medieval Sámi winter camps with turf huts).

Areas with high phosphate and low pH readings are also found at all sites. These might relate to other human and animal ac-

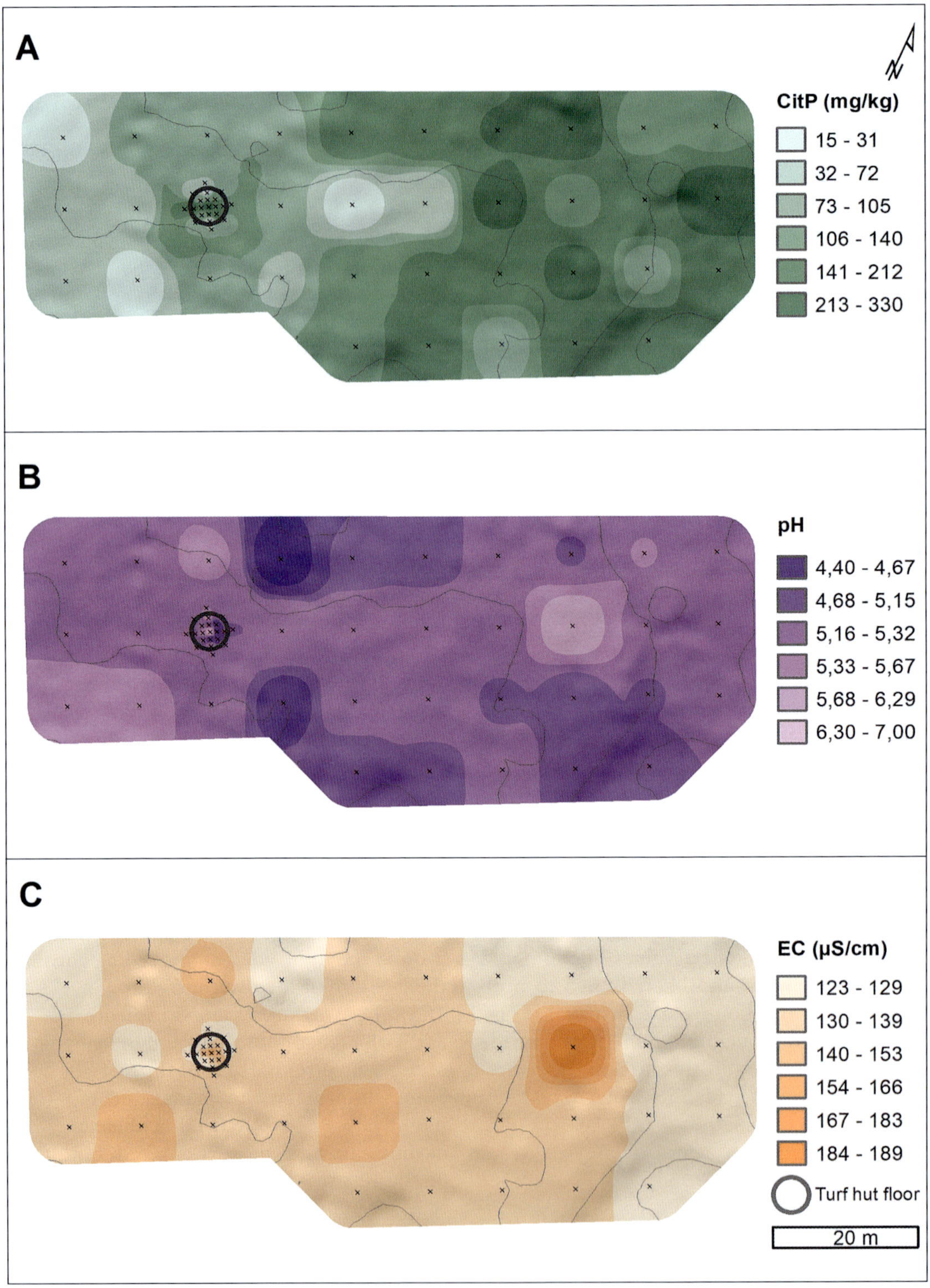

Figure 4. Measured geochemical values at Nuvkkágieddi: A. Phosphates. B. pH. C. Electric conductivity. Sampling locations with crosses, elevation contours at 0,5-metre intervals. Maps: Oula Seitsonen.

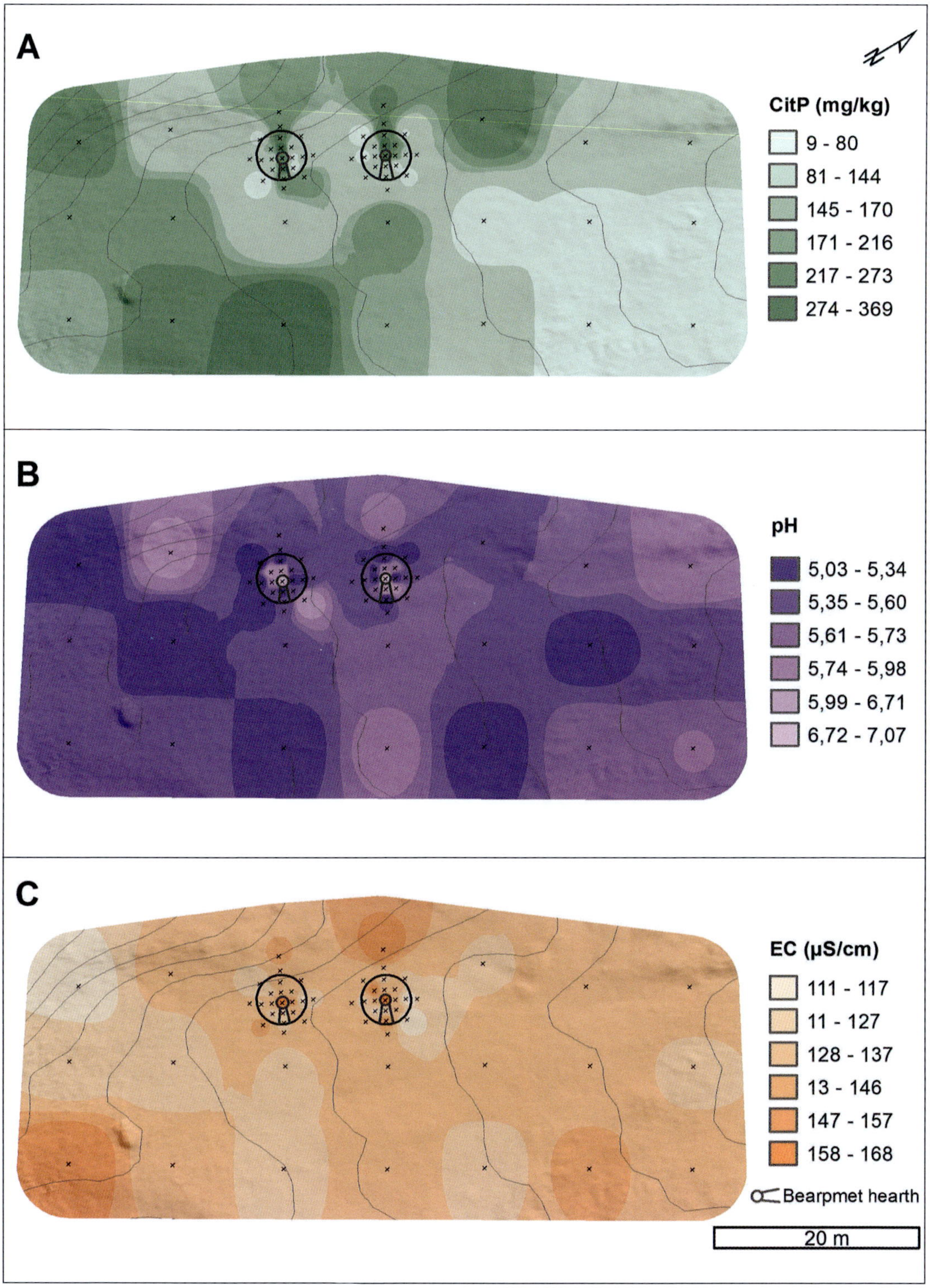

Figure 5. Measured geochemical values at Čáhkaljohka etelä 2: A. Phosphates. B. pH. C. Electric conductivity. Sampling locations with crosses, elevation contours at 0,5-metre intervals. Maps: Oula Seitsonen.

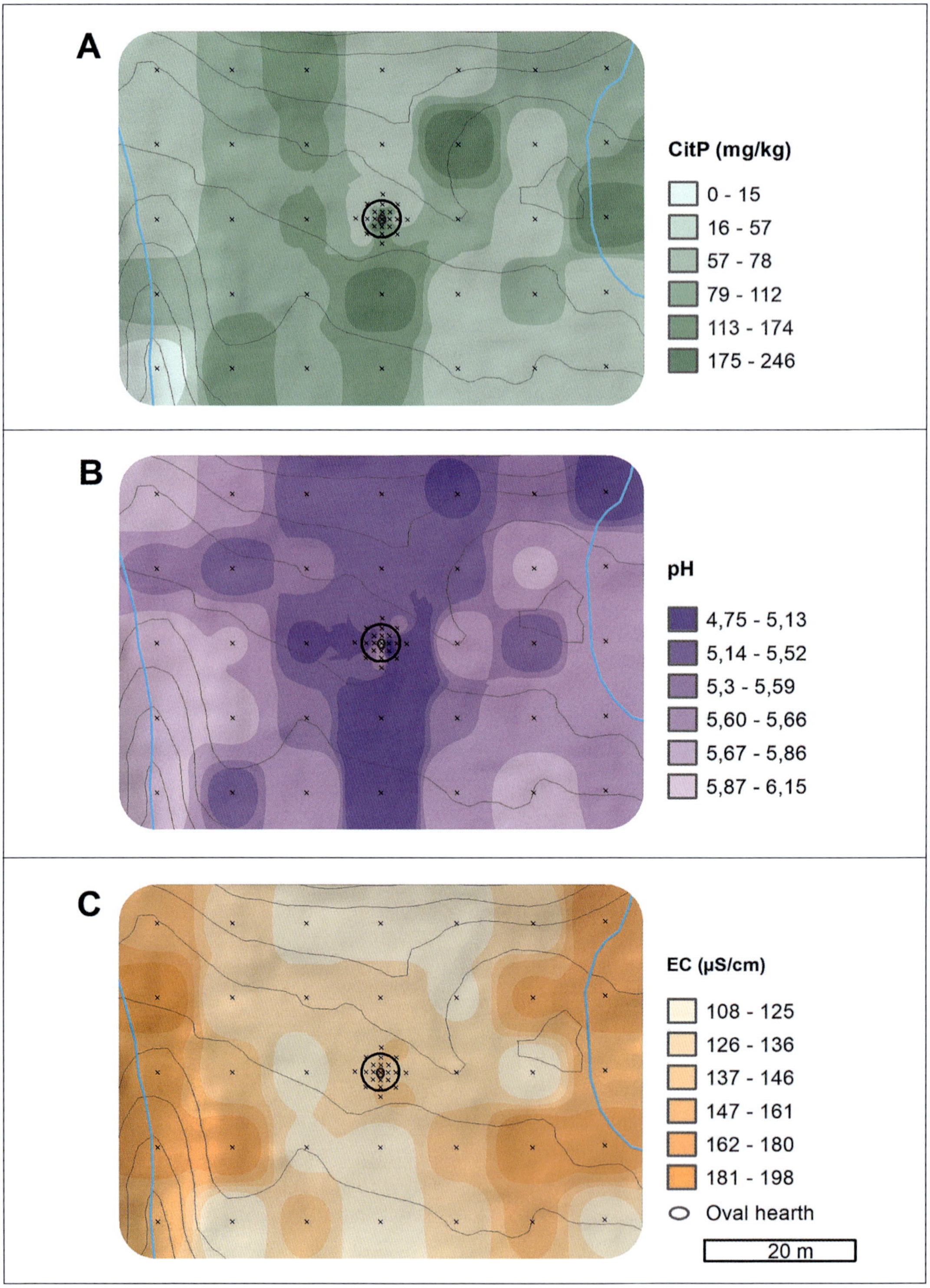

Figure 6. Measured geochemical values at Gahperusvárri itä 2: A. Phosphates. B. pH. C. Electric conductivity. Sampling locations with crosses, elevation contours at 0,5-metre intervals. Maps: Oula Seitsonen.

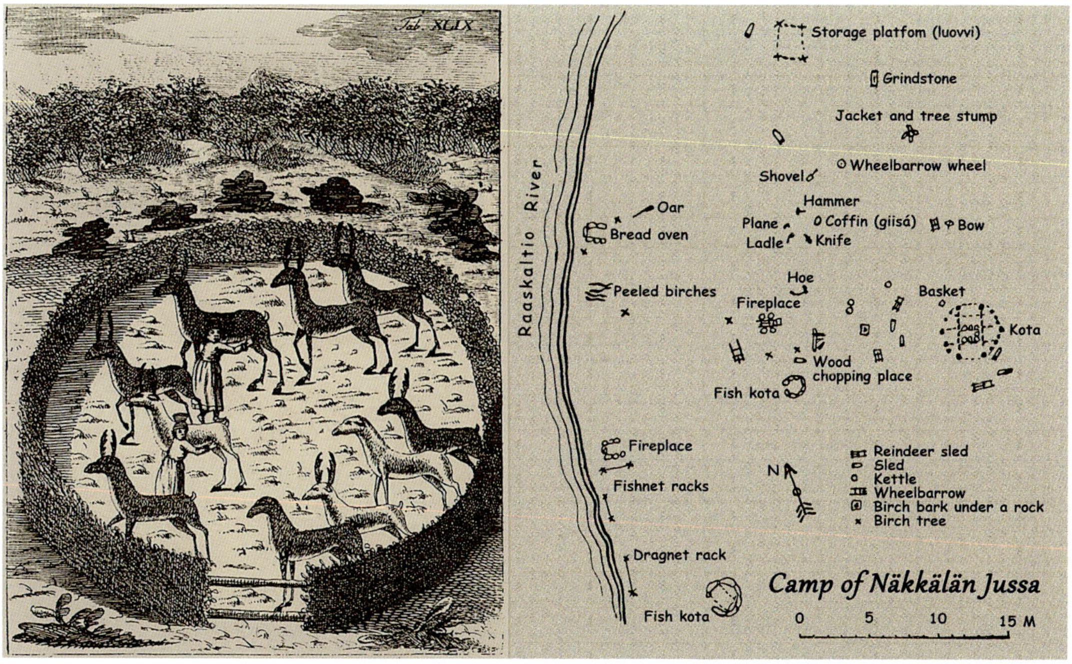

Figure 7. Left: Sámi women milking reindeer in a branch-built corral in the 18th century (after Leem, 1767: Plate XLIX, public domain), notice the ládjogahpir horn hats worn by the milking women: based on transgenerational memories these hats were forbidden by the Christian priests, alongside the noaidi drums, as material expressions of the traditional indigenous Sámi cosmology and folk religion (Pieski & Harlin 2020); Right: Mountain Sámi reindeer herding camp of Näkkälän Jussa (Jussa Näkkäläjärvi) in the early 20th century; note that the tent place (Kota) on the right shows the intra-dwelling division (modified after Paulaharju 1922: 94).

tivities than waste disposal. The high phosphate levels are indicative of increased human and animal activity (e.g. Karlsson 2006), and the lower pH could tell, for instance, of areas where animals have been gathered and where they have recurrently urinated, making the soils more acidic (Lungu & Dynoodt 2008). Based on the elevated phosphate values, Karlsson (2006) and Jerand and Linderholm (2019) have staked out potential reindeer gathering, and perhaps corralling, areas adjacent to the dwelling remains at the sites they have studied (Jerand & Linderholm 2019: 36; Karlsson 2006: 163). Such animal handling areas could have existed on both sides in front of the turf hut at Nuvkkágieddi (with low pH values) and in front of the dwellings at Čáhkaljohka and Gahperusvárri (with high phosphate levels and low pH).

Besides hearths, the highest EC readings come from the more fine-grained moist soils encountered along the edges of the Čáhkaljohka and Gahperusvárri sites, and thus likely relate more to natural processes than to human or animal activity. However, there is one spot with an exceptionally high EC value at Nuvkkágieddi about 40 metres northeast from the turf hut floor towards the shore of the adjacent lake, accompanied by a high pH reading. This could indicate the existence of an outside fireplace, although no charcoal or hearth stones were observed at the spot, or some metal deposit. Ethnographers have documented both open-air fireplaces and so-called fish-huts used for storing fish at Sámi camps, often towards the shoreline from the dwellings (Fig. 7) (e.g. Paulaharju 1922: 94). For example, a fish-hut might have been such a lightweight structure that it has left only high EC and pH values in the soil instead of visible surface traces. In addition, smoky open-air *suovas* fires without any structures have been used to help evade the swarms of mosquitoes and other insects plaguing the reindeer (e.g. Itkonen 1948b: 149, 165, 274). These kinds of fires could also show up as anomalously high EC values within the sites, although in this case there are no high phosphate readings that would link with wood ash deposition.

5 Intra-dwelling spatialities

The geoarchaeological properties of the soils within the turf hut floor and around the hearths were examined with a higher resolution to compare them against the often-used ethnographic analogies about the division of space inside the dwellings (e.g. Itkonen 1948a; Leem 1767; Rheen 1897). Here it suffices to note that these floor plans mirror in various ways the Sámi cosmologies and symbologies (e.g. Fossum 2006; Ränk 1949; Rydving 1995; Yates 1989).

The floor space was typically divided into five main portions along a mid-passage on the central axis of the dwelling from the doorway to the back (e.g. Ränk 1949). This could be further divided into altogether nine parts (Fig. 8; Itkonen 1948a: 184–186, 196; Mulk 1994: 205). In the ethnographic cases the floor plan was often visibly laid out with logs (Itkonen 1948a: 183), like some herders still do today (Seitsonen 2020), or with the stone lines visible at the *bearpmet* hearths which were used still in the early 20th century. In the archaeological cases where stones or other visible remains of the floor partitioning are absent, geochemical analyses have proven to be a fruitful approach for studying the use of space inside the dwellings (see Halinen 2009; Hedman & Olsen 2009; Karlsson 2006: 164). Based on these and spatial analyses, the floor plans inside Sámi dwellings appear to have been roughly analogous in different parts of Sápmi already during the Iron Age (e.g. Halinen 2016; 2019; Jerand et al. 2016; Nilsen 2015).

The five main parts of a Sámi dwelling are (Fig. 8):

1. *Uksa*: The doorway, entrance part of the hut where also firewood was stored

2. *Árran*: The fireplace in the centre of the dwelling, typically outlined with rocks, sometimes packed with smaller stones

3. *Boaššu*: The back of the hut, the kitchen area where food was prepared and kitchen utensils stored, and in pre-Christian times also the sacred space where the *noaidi* (shaman) drum and hunting equipment were stored; in part of Sápmi in pre-Christian times the kitchen area was limited to a constricted space between the *árran* and the *boaššu*, called *luops* (Huttunen 2015)

4–5. *Loaidu*: The side sections of the dwelling where people lived and slept on both sides of the mid-passage; the side sections can be further divided into three portions

Ethnographic source material often highlights the gendered divisions attached to the dwelling space (Mulk 1994: 205), although there are rather contradicting examples about the use of social space in different parts of Sápmi (Fellman 1906; Graan 1899; Itkonen 1948a: 184; Leem 1767; Ränk 1949; Rheen 1897; Schefferus 1673; Yates 1989). The gendered spatial restrictions of the social space mirror some widespread Pan-Arctic customs ranging from Europe to Siberia and the American Arctic, such as the male-related rear and female-related entrance areas (e.g. Haakansson 2001; Mulk 1994: 205; Nilsen 2015; Ränk 1949). In the Sámi dwellings, the sacred *boaššu* zone in the back is said to have been reserved for men and their activities, and the *boaššouksa*, a sacred backdoor of tents and turf huts mentioned in ethnography and used at places until the 19th century (Itkonen 1948a: 196), was also reserved for men and hunting-related activities, such

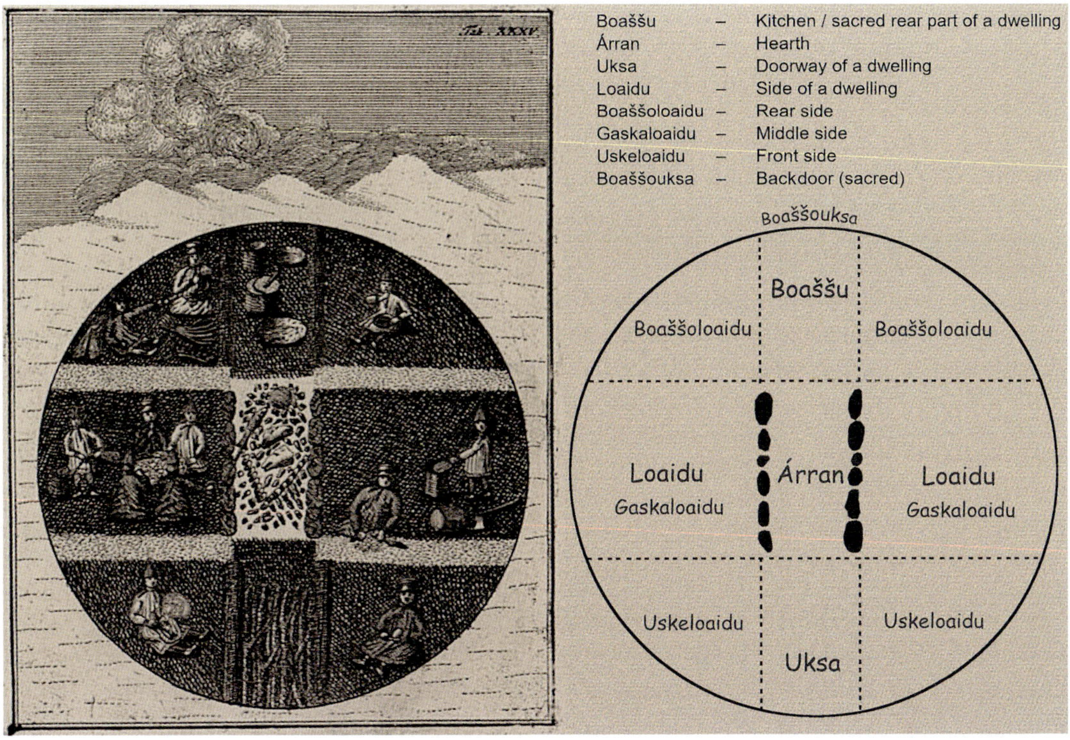

Figure 8. Left: The nine-part internal division of a Sámi dwelling in the 18th century (after Leem, 1767: Plate XXXV, public domain); Right: Schematic presentation of the different sectors inside a dwelling (based on Itkonen 1948a; Ränk 1949; Solbakk 2007).

as bringing in the wild game and fish (Halinen 2019; Mulk 1994: 205; Ränk 1949; Rheen 1897). In contrast, the frontal *uksa* area was related to the female deities and female activities, female utensils were stored there, and it was used for bringing in the domestic and dairy products (Mulk 1994: 205; Ränk 1949). However, spatial analyses of archaeological assemblages have shown that artefacts related to men's and women's activities are often found in different sectors than what the ethnographic sources direct to expect and suggest a fluid use of the social dwelling spaces (Halinen 2019; Inkiläinen 1999; Mulk 1994: 214).

The geoarchaeological values measured within the turf hut floor at Nuvkkágieddi and around the hearths at Čáhkaljohka and Gahperusvárri are illustrated in Figure 9. All of these show correlations to the schematic Sámi dwelling floor plans superimposed on them, centred on the hearth structures based on the ethnographic examples. Hearths and their immediate surroundings stand out with the highest readings in all the geochemical components. Spatial analysis suggests the presence of the *boaššouksa* backdoor, the sacred entrance known from the ethnographic sources but left out of use in the 19th century at the latest, at all the dwellings except perhaps at Čáhkaljohka (Fig. 9: B1–3).

At Nuvkkágieddi, high phosphate values, accompanied by low pH values, suggest entrances on both the southwest and northeast sides of the turf hut (Fig. 9: A1–3) (see Carpelan & Lavento 1996; Halinen 2009; Karlsson 2006: 142). A shallow depression is visible on the wall on the northeast side and could mark either the *uksa* or the *boaššouksa* (see Halinen 2019: 118). The area inside this depression and extending to both *loaidu* has low pH values, perhaps indicative of human activities making the soils more acidic. Low phosphate values were recorded on the turf walls, whereas outside the hut is surrounded on three sides by higher readings. This might relate to waste management, such as the dumping of bone waste along the walls of the turf hut (Halinen 2019: 114).

At Čáhkaljohka, based on the spatiality of the geochemical readings, the *uksa* seems to have been directed south, and there might have been no backdoor. The area around the hearth has higher phosphates in all the floor plan sectors, the highest pH readings are found in the *árran* and *uksa* sections and to some extent in the *boaššu*, and high EC values are found along the mid-passage (Fig. 9: B1–3). The low pH measured in the eastern loaidu might relate to human activities. Then again, the high pH and EC immediately outside the assumed wall-line on the northeast side could originate from the dumping of bone waste and burnt material behind the dwelling (Itkonen 1948a: 185), although the phosphate values are lower than inside.

In the *bearpmet* hearths at Gahperusvárri, the *uksabearpmet* stone lines mark that the entrances of the tents faced southeast. The phosphate and pH values of both hearths suggest the presence of the *boaššouksa* on the rear (Fig. 9C1–3; 9D1–3). The *Boaššu* and *árran* of both hearths are characterised by high pH and phosphate values, as is also the area on the right side of the entrance of hearth 1, probably indicating the handling and disposal of bones and other waste in these areas. Contrary to the entrances of all the other studied structures, the *uksa* area of the Gahperusvárri hearth 2 has low phosphate values. The spatiality of the geochemical readings from the Gahperusvárri hearth 1 conforms most closely to the stereotypical pattern that might be expected based on the ethnographic examples: high phosphate values along the mid-passage and extending outside through the doorways in the front and the back, low phosphate values in both *loaidu*, high pH readings in the *árran* and *boaššu*, and high EC values in the *árran* (see Karlsson 2006: 142).

6 Transspecies lifeworlds of the historical reindeer herder camps

The materialities of soil at the studied Sámi reindeer herder camps has been shaped by the transspecies human-animal interaction in their overlapping embodied lifeworlds. Lifeworld has been described by geographer David Seamon (2017: 248) as the 'everyday realm of experiences, actions, and meanings typically taken for granted…' This enables the animal or human subject's perceptual and pre-cognitive bodily immersion in and intertwining with their surrounding world (Russell 2016; Seamon 2013). The transspecies spatial theory, as advocated by geographers Andrea Bolla and Alice Hovorka (2012), perceives animals as another social group that engages in an active and dynamic relationship with people (also Tuan 1984; Wolch et al. 1995). The human-animal sociospatial practices are always bidirectional, and the animals may exercise their own power and agency over the humans, e.g. by accepting or averting their care and handling (Hovorka 2020). This is well in line with the Sámi environmental perception and cosmology that approaches the landscape as a relational, inclusive and cognitively controlled unity that evades the 'Western' dualistic oppositions of nature and culture (e.g. Ruotsala 2002: 331; Seitsonen 2020: 16). Sámi pastoralists perceive the landscape in its entirety as their extended home; like one of our informants expressed: 'Home is all our fells.'

The studied sites form one small portion of the vast meshwork of relations that shroud the Sámi pastoralists' home range and form their various taskscapes on the tundra (Ingold 1993; 2011). Archaeologist Tiina Äikäs (2017: 217) has noted that Sámi perceive their landscape through motion, activity and participation, and these 'activities produce different ways of experiencing and structuring the world' into multiple overlapping and relationally entangled taskscapes, or *meahcci* (see Joks et al. 2020; Mazzullo & Ingold 2008; Schanche 2002). *Meahcci* is a Northern Sámi word that signifies the areas where

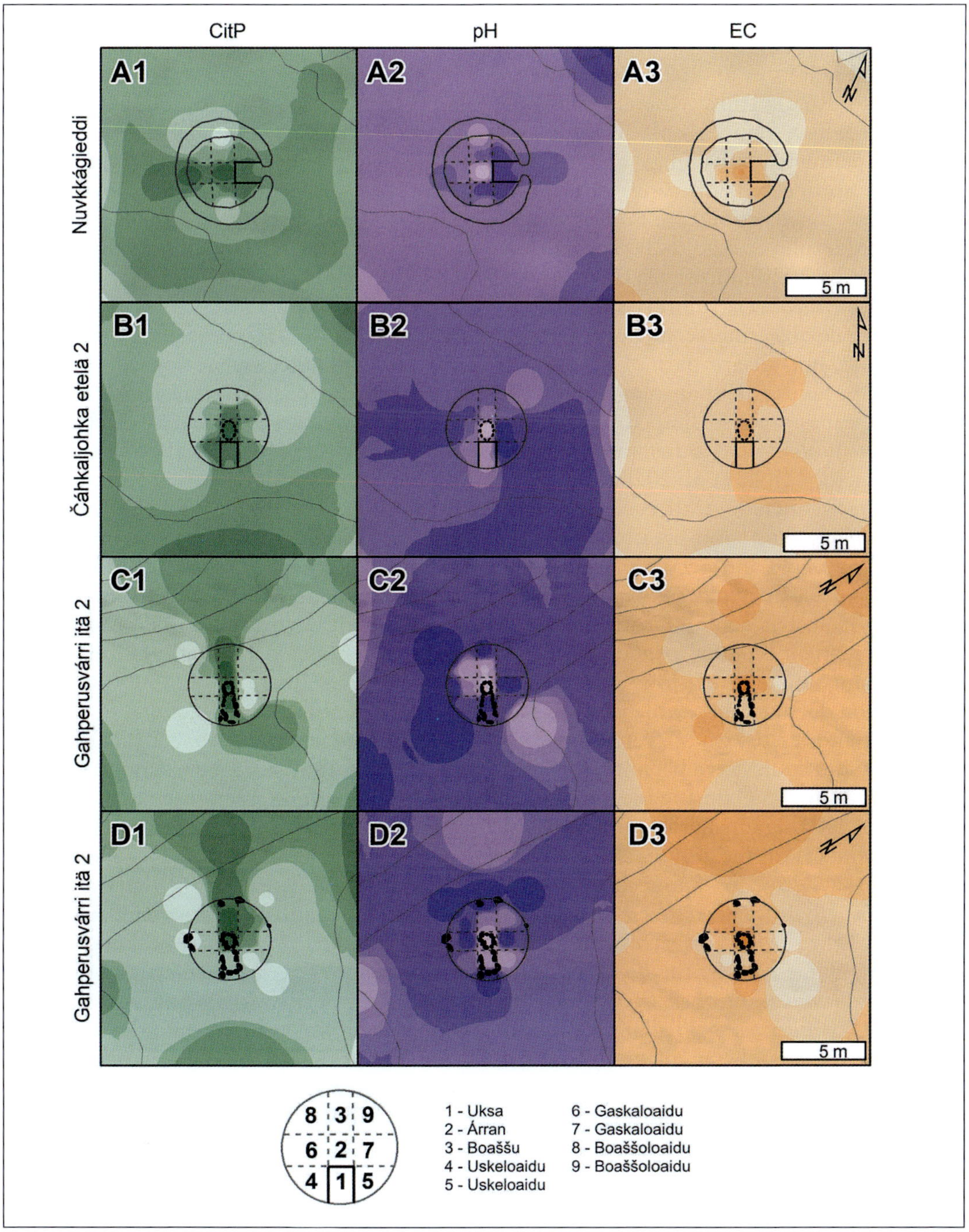

Figure 9. Intra-dwelling geoarchaeological values and suggested internal divisions: A1–3. Turf hut floor at Nuvkkágieddi, shallow depression on the wall to the right, possibly an entrance. B1–3. Oval hearth at Čáhkaljohka etelä 2, entrance likely towards the bottom. C1–3. Bearpmet hearth 1 at Gahperusvárri itä 2, entrance towards the bottom. D1–3. Bearpmet hearth 2 at Gahperusvárri itä 2, entrance towards the bottom. See Fig. 4–6 for the sampling locations and legends of the phosphate, pH and EC values, and Fig. 8 and the text for further explanation of the intra-dwelling division. Elevation contours at 0,5-metre intervals. Map: Oula Seitsonen.

varying spatiotemporal tasks are performed but where people do not live (Tervaniemi & Magga 2019; Joks et al. 2020; Seitsonen & Moshenska 2020). Thus, *Meahcci* contrasts with *báiki*, which signifies a place where seasonal or permanent dwellings and storage structures are located. *Meahcci* starts from where these constructions are no longer observable (Mazzullo & Ingold 2008; Seitsonen & Moshenska 2020). However, the boundaries between *meahccit* (plural) and *báikkit* (plural) are flexible, intersecting and seasonally shifting, but all are part of a cohesive herder home landscape (Länsman 2004: 99). Sámi taskscapes are also shaped by the '*unfolding encounters* with other more or less powerful actors' (Joks et al. 2020: 308, original emphasis) including all ecological life, both breathing and non-breathing things, as well as invisible and occult entities, such as sacred places, story places or visions, and mirror an inclusive, communicative and negotiating 'sentient ecology' (Anderson 2000: 116–117; Ingold 2000: 25; Joks et al. 2020).

Based on documentary evidence from the 18th–19th centuries, the *báikkit* represented by our study sites could have been inhabited in the spring (May–June) during the northward migrations, and again in the autumn (August–September) on the way back south (Itkonen 1948b: 225–226). Snow cover lasts in this area usually from September–October to May–June. The sites would have been immediately surrounded e.g. by the *boazomeahcci* pasturelands, *muorrameahcci* firewood gathering areas, *guollemeahcci* fishing lakes and *luomemeahcci* cloudberry picking swamps (Joks et al. 2020). Other *meahcci* taskscapes included in the herders' annual sphere would at least have been their winter pastures in the forested *vuovdi* lands nearly 100 kilometres to the south, their summer camps and pastures on the Norwegian coast over 50 km to the north (before the border closures in the mid-19th century), and the spring calving and autumn rutting grounds on the fells in between (e.g. Itkonen 1948b: 225–226). Other, overlapping *meahcci* would have included the fishing, hunting and gathering grounds of the different families and *siidas* based on the *vuoigatvuohta* ancestral rights defined by the Sámi common law (e.g. Helander 2000; Joks et al. 2020; Schanche 2002).

Owing to the transitional character of the studied *báikkit*, it is not surprising that their phosphate readings are much lower than those from the Iron Age and medieval winter camps that were likely inhabited throughout the long winter months (Carpelan & Lavento 1996; Halinen 2009; Jerand et al. 2016; Karlsson 2006). The archaeological remains and associated geoarchaeological values result from the past, overlapping temporalities of the human and animal place-ballets, the 'interaction of individual bodily routines rooted in a particular environment' (Seamon 2013: 206), collapsed into the landscape (Mlekuž 2014). These intertwining place-ballets would have been shaped and directed by topography, soils, vegetation cover and the built environment, characterised by tents, turf huts, corrals, storage racks, garbage dumps and so on (Fig. 7). At the same time, the mobilities and actions of humans and animals mutually shaped all of these. The hearths and turf huts, and any other standing structures such as storage racks or corrals, visible from the distance in the open landscapes, could have acted as anchors for the spatial configuration of the sites. Both human and animal activities and mobilities likely took place in relation to these.

Based on the geochemical analyses, it appears that the waste management at the studied sites was analogous to the excavated Sámi localities, where the bone waste was typically deposited along the walls and especially behind the dwellings, and inside the dwellings at the *boaššu* and *árran* zones (e.g. Carpelan 1992; Halinen 2009; 2019: 114; Hedman & Olsen 2009; Olsen 2019). This is also reminiscent of the ethnographic accounts about waste management (Itkonen 1948a: 185). Inside the dwellings, the spatial configuration of space and place-ballets of the inhabitants were guided by the cultural rules and customary activities that then shaped the materialities of soil. All the other studied dwellings except the hearth at Čáhkaljohka clearly suggest the presence of a *boaššouksa* backdoor (Itkonen 1948a: 196; Ränk 1949).

At all the sites, the phosphate and pH readings suggest that reindeer gathering and handling took place in front of the dwellings, within 20–40 metres from them. The dairy economy formed an important part of the Mountain Sámi subsistence as recently as the early 20th century (Itkonen 1948a: 273), and these areas might represent the daily reindeer milking zones. In the early 20th century, milking places were typically near the dwellings (Itkonen 1948b: 148) as also suggested by Karlsson's (2006: 163) archaeological studies, either in the open or inside branch-built milking corrals (*gárdi*) (Fig. 7) (Itkonen 1948b: 143, 148, 274; Karlsson 2006: 163). According to ethnographer Toivo I. Itkonen (1948a: 273–274), the milking of reindeer was in earlier times done primarily by men, though women could also do it, but later it became increasingly a female task.

It is known from the ethnographic sources that the daily reindeer mobilities and their place-ballets were at least sometimes directed by the initiative of the animals themselves, e.g. by their grazing needs and the mosquitoes and other bugs pestering them (Paulaharju 1922: 99). The reindeer moved on their own incentive in the warmest time of the day from their highland pastures to the camps, where they knew that the smoky *suovas* fires lit by the herders offered them respite from the swarms of insects, and also to the windy capes, followed and, if needed, guided by the herders and their dogs (Paulaharju 1922: 98–99). At the camps they were milked, and in the evening, when the worst insect pestilence eased, the animals started again moving to their pastures on the surrounding fells followed by the dogs and herders for the night (Paulaharju 1922: 100–103). Thus, the transspecies spatialities of the herder camps included at least two more major actors besides reindeer and humans, namely the reindeer dog, *boazobeana*, and the various flying insects. Both have their own role in directing and affecting the reindeer and human mobilities and behaviour. The herding dogs contributed directly to the geochemical signals at the sites, but the insect contribution to the archaeological record is indirect and mostly invisible. However, the insects have even received praise for their participation in herding by the reindeer herders, like one joik (traditional Sámi song) declares:

> 'Mosquito drives the reindeer to the fells, herds the reindeer together out from the valleys. You wouldn't get along with the reindeer, if no mosquitoes existed. It is a small creature, but in front of it a large herd will escape.' (Paulaharju 1922: 99, our translation)

7 Conclusion

The geoarchaeological analyses of the three studied sites emphasise the high potential that soil studies have in examining the spatialities, use length and intensity of the historical Sámi reindeer herding camps. The previous soil-based analyses of Sámi sites (e.g. Carpelan & Lavento 1996; Halinen 2009; Karlsson 2006; Nuñez 1977; Tolonen 2013) have highlighted the suitability of different geochemical and geophysical variables for intrasite studies. In this study we analysed the phosphate values, which have been most commonly used in the studies, pH, which is a more rarely analysed property, and electric conductivity, which is a new variable that has not been previously assessed in a Sámi context. All the studied elements illustrate different aspects of the spatial configurations of the campsites. Based on comparisons with the ethnographic examples (e.g. Itkonen 1948a; Paulaharju 1922; Ränk 1949) and excavated localities (e.g. Jerand & Linderholm 2019; Jerand et al. 2016; Halinen 2019), our analyses illustrate the culturally demarcated floor plans of the dwellings, waste management and potential reindeer handling areas at the studied sites. These allow important advancements for the study of intrasite patterns and use of space at the Sámi pastoralist campsites in the little studied Gilbbesjávri region, and also more widely in Sápmi. In the future, in-situ field-based or ex-situ laboratory-based portable X-ray Fluorescence (pXRF) analyses could offer a cost-effective way forward for analysing multiple elements from the soils (see Hunt & Speakman 2014; Williams et al. 2020).

Materialities of soil at the studied sites were shaped by the transspecies human-animal interplay (Hovorka, 2020), which included at least the Sámi, their reindeer and herding dogs, and, indirectly, different bugs pestering both humans and animals. The reindeer agency and initiative in the habitual place-ballets that shaped the soil geoarchaeologies is documented in the ethnographic literature and was motivated e.g. by the smoky *suovas* fires lit by the herders to offer respite from the swarms of insects plaguing their animals. Studying the overlapping and intertwining lifeworlds of the different social groups, including human and animal actors (see Äikäs et al. 2021; Nyyssönen & Salmi 2013; Salmi & Heino 2019) such as reindeer, dogs, mosquitoes and parasitic flies (Kynkäänniemi 2020), allows approaching and interpreting the spatialities of reindeer herder camps more holistically.

Acknowledgements

We want to thank all the locals in Gilbbesjávri who have assisted in our studies, especially the reindeer master Juha Tornensis of the Käsivarsi paliskunta and Antti Ohenoja of the National Board of Forestry, the Sámi Parliament and Sámi Museum Siida for the continued cooperation, and the reviewers and editors for their comments on an earlier draft. This research is part of the project Domestication in Action (Academy of Finland grant 308322; European Research Council grant 2017 756431).

References

Unpublished sources

Halinen, P. 2008a. Utsjoki 212 Pulmankijärvi Kalddasjohka 4 (1000001231) kaivaus, 27.-29.8.2008. Research report, National Heritage Agency, Helsinki.

Halinen, P. 2008b. Utsjoki 362 Jeagelveijohka 1 (1000001211) kaivaus, 21.-24.7.2008. Research report, National Heritage Agency, Helsinki.

Näkkäläjärvi, K. 2013. Jauristunturin poropaimentolaisuus: kulttuurin kehitys ja tietojärjestelmä vuosina 1930-1995. Ph.D. thesis, Anthropology. University of Oulu, Oulu.

Nilsen, A. C. 2015. The Slettnes type: Proto-Sámi dwellings in Northern Norway 0–1050 AD. M.A. thesis, University of Tromsø, Tromsø.

Seitsonen, O. 2021b. Enontekiö. Esihistoriallisen ja historiallisen ajan saamelaisten poronhoitokohteiden kartoitus sekä ajoitus- ja maaperänäytteiden otto 21.7.–15.8.2019. Research report, University of Oulu, Oulu.

Tolonen, S. 2013. Uhraamisen sijoittuminen seidoilla. Geokemiallisia ja fysikaalisia havaintoja ja tilallista tulkintaa Enontekiön Näkkälän, Utsjoen Sieiddakeädgin, Muonion Kirkkopahdan ja Muonion Porviniemen seitakohteilta. M.A. thesis, University of Oulu, Oulu.

Literature

Äikäs, T. 2015. *From boulders to fells: sacred places in the Sámi ritual landscape.* Monographs of the Archaeological Society of Finland 5.

Äikäs, T. 2017. Sámi sacred places in ritual taskscapes. In U. Rajala & P. Mills (eds.) *Forms of Dwelling: 20 Years of Taskscapes in Archaeology*: 215–235.

Äikäs, T., Seitsonen, O. & Salmi, A.-K. 2021. Kolari marketplace in a multicultural landscape between Sámi, Swedes and Finns. ***META*** 2021: 125–146.

Anderson, D. G. 2000. *Identity and Ecology in Arctic Siberia: The Number One Reindeer Brigade.* New York: Oxford University Press.

Bolla, A. K. & Hovorka, A. J. 2012. Placing Wild Animals in Botswana: Engaging Geography's Transspecies Spatial Theory. *HUMaNIMALIA* 3 (2): 56–82.

Bronk, Ramsey, C. 2017. Methods for Summarizing Radiocarbon Datasets. *Radiocarbon* 59 (2): 1809–1833.

Carpelan, C. 1992. Juikenttä – näkökulma saamelaiseen yhteiskuntaan. *Studia Historica Septentrionalia* 21: 34–44.

Carpelan, C. & Lavento, M. 1996. Soil phosphorus survey at subrecent Saami winter village sites near Inari, Finnish Lapland. A preliminary report. In V. Mejdahl & P. Siemen (eds.) *Proceedings from the 6th Nordic Conference on the Application of Scientific Methods in Archaeology. Esbjerg 1993*: 97–107. Arkæologiske Rapporter 1.

Égüez, N. & Makarewicz, C. A. 2018. Carbon isotope ratios of plant n-alkanes and microstratigraphy analyses of dung accumulations in a pastoral nomadic winter campsite (Eastern Mongolia). *Ethnoarchaeology. Journal of Archaeological, Ethnographic and Experimental Studies* 10 (2): 141–158.

Fellman, J. 1906. *Anteckningar under min vistelse i Lappmarken* III. Stockholm: Kungl. boktryckeriet.

Fenger-Nielsen, R., Hollesen, J., Matthiesen, H., Andersen, E. A. S., Westergaard-Nielsen, A., Harmsen, H., Michelsen, A. & Elberling, B. 2019. Footprints from the past: The influence of past human activities on vegetation and soil across five archaeological sites in Greenland. *Science of the Total Environment* 654 (1): 895–905.

Fossum, B. 2006. *Förfädernas land. En arkeologisk studie av rituella lämningar i Sápmi, 300 f. Kr.–1600 e. Kr.* Studia Archaeologica Universitatis Umensis 22.

Graan, O. 1899. *Relationen, eller en fulkomblig beskrifting om Lapparnas ursprung, så wähl some hela dheras lefwerne förehållande. Nertecknad 1672.* Bidrag till kännedom om de svenska landsmålen och svenskt folkliv 17:2.

Haakansson, S. D. 2001. *Ethnoarchaeology of the Yamal Nenets: Utilizing emic and etic evidence in the interpretation of archaeological residues.* Cambridge: Harvard University.

Halinen, P. 2009. Change and Continuity of Saami dwellings and dwelling sites from the Late Iron Age to the 18th century. In T. Äikäs (ed.) *Máttut–máddagat: The Roots of Saami Ethnicities, Societies and Spaces/Places*: 100–115. Publications of the Giellagas Institute 12.

Halinen, P. 2016. The Inland Sámi Societies of Northern Fennoscandia during the Late iron Age and Early Medieval Period: An Archaeological Approach. In P. Uino & K. Nordqvist (eds.) *New Sites, New Methods*: 160–174. Iskos 21.

Halinen, P. 2019. Inari Nukkumajoki 5, the Excavated History of a Sámi Winter Village. In P. Halinen & B. Olsen (eds.) *In Search of Hearths. A Book in Memory of Sven-Donald Hedman. Proceedings of the Seminar Dedicated the Memory of Sven-Donald Hedman, University of Tromsø – The Arctic University of Norway, March 4, 2016*: 96–122. Iskos 22.

Halinen, P., Hedman, S.-D. & Olsen, B. 2013. Hunters in Transition: Sámi Hearth Row Sites, Reindeer Economies and the Organization of Domestic Space 800-1300 A.D. In D.G. Anderson, R. Wishart & V. Vaté (eds.) *About the Hearth – Perspectives on the Home, Hearth and Household in the Circumpolar North*: 152–182.

Hansen, L. I. & Olsen, B. 2014. *Hunters in Transition. An Outline of Early Sámi History*. Leiden: Brill.

Hedman, S.-D. & Olsen, B. 2009. Transition and Order: a Study of Sámi Rectangular Hearths in Pasvik, Arctic Norway. *Fennoscandia archaeologica* XXVI: 3–22.

Heikkinen, H., Jokinen, M., Valkeapää, O. A. & Helle, T. 2005. Poronhoidon historia Käsivarren Lapissa ja Mallalla. In M. Jokinen (ed.) *Poronhoidon ja suojelun vaikutukset Mallan luonnonpuistossa*: 14–24. Metsäntutkimuslaitoksen tiedonantoja 941.

Helander, E. 2000. Saamelainen maailmankuva ja luontosuhde. In I. Seurujärvi-Kari (Ed.) *Beaivvi mánát. Saamelaisten juuret ja nykyaika*: 171–182. Tietolipas 164.

Hovorka, A. J. 2020. Animal geographies. In A. Kobayashi (ed.) *International Encyclopedia of Human Geography*: 127–132.

Hunt, A. & Speakman, R. J. 2014. Portable XRF analysis of archaeological sediments and ceramics. *Journal of Archaeological Science* 53: 626-638.

Husserl, E. 1970 [1936]. *The Crisis of European Sciences and Transcendental Phenomenology: An Introduction to Phenomenological Philosophy*. Evanston: Northwestern University Press.

Huttunen, M. 2015. Salvoskota, hirsitupa ja hirsitalo – inarinsaamelaisten vanhat hirsiasumukset. In P. Magga & E. Ojanlatva (eds.) *Ealli biras. Elävä kulttuuriympäristö. Saamelainen kulttuuriympäristöohjelma*, 98–109.

Ingold, T. 1993. The temporality of the landscape. *World Archaeology* 25 (2): 152–174.

Ingold, T. 2000. *The Perception of the Environment: Essays on Livelihood, Dwelling and Skill*. London: Routledge.

Ingold, T. 2011. *Being alive: Essays on movement, knowledge and description*. New York: Routledge.

Inkiläinen, M. 1999. Nukkumajoki 2 -saamelaiskylä sukupuoliarkeologisena tutkimuskohteena. *Muinaistutkija* 3/1999: 2–14.

Itkonen, T. I. 1948a. *Suomen lappalaiset vuoteen 1945* I. Porvoo: WSOY.

Itkonen, T. I. 1948b. *Suomen lappalaiset vuoteen 1945* II. Porvoo: WSOY.

Jerand, P., Linderholm, J., Hedman, S.-D. & Olsen, B. 2016. Spatial perspectives on hearth row site organisation in Northern Fennoscandia through the analysis of soil phosphate content. *Journal of Archaeological Science: Reports* 5: 361–373.

Jerand, P. & Linderholm, J. 2019. Possible Traces of Reindeer Corralling at a Hearth-Row Site from 1000–1300 CE in Northern Norway. *In P. Halinen & B. Olsen (eds.) In Search of Hearths. A Book in Memory of Sven-Donald Hedman. Proceedings of the Seminar Dedicated the Memory of Sven-Donald Hedman, University of Tromsø – The Arctic University of Norway, March 4, 2016*: 31–39. Iskos 22.

Joks, S., Østmo, L. & Law, J. 2020. Verbing meahcci: Living Sámi lands. *The Sociological Review*. 68 (2): 305-321.

Karlsson, N. 2006. *Bosättning och resursutnyttjande. Miljöarkeologiska studier av boplatser med härder inom perioden 600–1900 e. Kr inom skogssamiskt område.* Studia Archaeologica Universitatis Umensis 21.

Kynkäänniemi, S.-M. 2020. *The relationship between the reindeer (Rangifer tarandus tarandus) and the ectoparasitic deer ked (Lipoptena cervi): reindeer welfare aspects*. Acta Universitatis Ouluensis A Scientiae Rerum Naturalium 741.

Länsman, A.-S. 2004. *Väärtisuhteet Lapin matkailussa Kulttuurianalyysi suomalaisten ja saamelaisten kohtaamisesta*. Inari: Kustannus-Puntsi.

Lavento, M. 2003. Maannos, maaperä ja niiden tutkimuksen menetelmät; arkeologin näkökulma. In P. Pesonen & E. Raike (eds.) *Luonnontieteelliset menetelmät ja GIS arkeologiassa*: 46–60. Arkeologipäivät 2003.

Leem, K. 1767. *Beskrivelse over Finmarkens Lapper, deres tungemaal, levemaade og forrige afgudsdyrkelse*. Kiøbenhavn: Kongl. Wäysenhuses Bogtrykkerie af G. G. Salikath.

Linderholm, J. 2010. *The soil as a source material in archaeology: Theoretical considerations and pragmatic applications*. Archaeology and Environment 25.

Lungu, O. I. M. & Dynoodt, R. F. P. 2008. Acidification from Long-term Use of Urea and its Effect on Selected Soil Properties. *African Journal of Food Agriculture Nutrition and Development* 8 (1): 63–76.

Mazzullo, N., and T. Ingold 2008. "Being along: Place, time and movement among Sámi people. In J. O. Baerenholdt & B. Granaas (eds.) *Mobility and Place. Enacting Northern European Peripheries*: 27–38.

Mlekuž, D. 2014. Exploring the topography of movement. In S. Polla, & P. Verhagen (eds.) *Computational Approaches to the Study of Movement in Archaeology Theory, Practice and Interpretation of Factors and Effects of Long-term Landscape Formation and Transformation*: 5–21.

Mulk, I.-M. 1994. *Sirkas – ett samiskt fångstsamhälle i förändring Kr.f.–1600 e.Kr*. Studia archaeologica universitatis umensis 6.

Nuñez, M. 1977. *Archaeology through Soil Chemical Analysis: an Evaluation*. Helsingin yliopiston arkeologian laitos, moniste 14.

Nykänen, T. & Valkeapää, L. 2016. Etniset poronhoitajat. Enontekiön porosaamelaisten ryhmämuotoisuus. In T. Nykänen & L. Valkeapää (eds.) *Kilpisjärven poliittinen luonto*: 160–198. Suomalaisen kirjallisuuden seuran toimituksia 1422.

Nyyssönen, J. & Salmi, A.-K. 2013. Towards a Multiangled Study of Reindeer Agency, Overlapping Environments, and Human–Animal Relationships. *Arctic Anthropology* 50 (2): 40–51.

Olsen, B. 2019. Brodkorpneset and Steintjørna: Two hearth-row sites in Pasvik, Arctic Norway. In P. Halinen & B. Olsen (eds.) *In Search of Hearths. A Book in Memory of Sven-Donald Hedman. Proceedings of the Seminar Dedicated the Memory of Sven-Donald Hedman, University of Tromsø – The Arctic University of Norway, March 4, 2016*: 9–30. Iskos 22.

Paulaharju, S. 1922. *Lapin muisteluksia*. Helsinki: Kirja.

Pieski, O. & Harlin, E.-K. 2020. *The Ládjogahpir. Máttaráhkuid gábagahpir. The foremothers' hat of pride*. Kárášjohka: Davvi Girji.

Ränk, G. 1949. Grundprinciper för disponeringen av utrymmet i de lapska kåtorna och gammerna. *Folkliv. Acta ethnologica et folkloristica Europaea TOM XII–XIII*: 87–111.

Reimer, P. J., Bard, E., Bayliss, A., Beck, J. W., Blackwell, P. G., Bronk Ramsey, C., Grootes, P. M., Guilderson, T. P., Haflidason, H., Hajdas, I., Hatt'e, C., Heaton, T. J., Hoffmann, D. L., Hogg, A. G., Hughen, K. A., Kaiser, K. F., Kromer, B., Manning, S. W., Niu, M., Reimer, R. W., Richards, D. A., Scott, E. M., Southon, J. R., Staff, R. A., Turney, C. S.M. & van der Plicht, J. 2013. IntCal13 and Marine13 Radiocarbon Age Calibration Curves 0-50,000 Years cal BP. *Radiocarbon* 55 (4): 1869–1887.

Rheen, S. 1897. *En kortt relation on Lapparnas lefwarne och sedher wijdskiepellsser, sampt i många stycken grofwe wildfarellsser. Nertecknad 1671*. Bidrag till kännedom om de svenska landsmålen och svenskt folkliv 17:1.

Ruotsala, H. 2002. *Muuttuvat palkiset. Elo, työ ja ympäristö Kittilän Kyrön paliskunnassa ja Kuolan Luujärven poronhoitokollektiiveissa vuosina 1930–1995*. Kansatieteellinen Arkisto 49.

Russell, J. 2016. Animal narrativity. Engaging with Story in a More-Than-Human World. In J. Castricano & J. Corman (eds.) *Animal Subjects* 2.0: 145–174.

Rydving, H., 1995. *The End of Drum-Time. Religious Change among the Lule Saami, 1670s–1740s*. Acta Universitatis Upsaliensis Historia Religionum, 12.

Salmi, A.-K. & Heino, M. 2019. Tangled worlds: The Swedish, the Sámi and the reindeer. *International Journal of Historical Archaeology* 23: 260–282.

Schanche, A. 2002. Meahcci – den samiske utmark. *Dieđut* 1: 156–171.

Schefferus, J. 1673. *Lapponia : id est, regionis Lapponum et gentis nova et verissima descriptio. In qua multa de origine, superstitione, sacris magicis, victu, cultu, negotiis Lapponum, item Animalium, metallorumque indole, quæ in terris eorum proveniunt, hactenus incognita Produntur, & eiconibus adjectis cum cura illustrantur*. Francofvrti: Ex Officina Christiani Wolffii. Typis Joannis Andreæ.

Seamon, D. 1979. *A Geography of the Lifeworld. Movement, Rest and Encounter*. London: Croom Helm.

Seamon, D. 2013. Place Attachment and Phenomenology: The Synergistic Dynamism of Place. In L.C. Manzo & P. Devine-Wright (eds.) *Place Attachment: Advances in Theory, Methods, and Applications*: 11–22.

Seamon, D. 2017. Architecture, Place, and Phenomenology: Buildings as Lifeworlds, Atmospheres, and Environmental Wholes. In J. Donohoe (ed.) *Place and Phenomenology*: 247–264.

Seitsonen, O. 2020. *Archaeologies of Hitler's Arctic War. Heritage of the Second World War German Military Presence in Finnish Lapland*. Oxon: Routledge.

Seitsonen, O. 2021a. 'We are always coming and going, like migratory birds': Diachronic changes in the seasonal settlement of Sámi reindeer herders in the Lake Gilbbesjávri region, northwestern Sápmi, ca. 700–1950 AD. In P. Dixon & C. Theune (eds.) *Seasonal Settlement in the Medieval and Early Modern Countryside*.

Seitsonen, O. & Moshenska, G. 2020. Who owns the 'Wilderness'? Indigenous Second World War Landscapes in Sápmi, Finnish Lapland. In N. Saunders & P. Cornish (eds.) *Conflict Landscapes: Materiality and Meaning in Contested Places*.

Solbakk, A. 2007. *Sápmi. Sameland. Samenes historie fram til 1751*. Kárášjohka: Davvi Girji.

Tervaniemi, S. E. & Magga, P. 2019. Belonging to Sápmi: the Sámi conceptions of home and home region. In T. H. Eriksen, S. Valkonen & J. Valkonen (eds.) *Knowing from the indigenous North: Sámi approaches to history, politics and belonging*: 75–90.

Trinks, I. 2014. Nondestructive Subsurface Mapping in Field Archaeology. In Smith, C. (ed.) *Encyclopedia of Global Archaeology*: 5297–5304. New York: Springer.

Tuan, Y.-F. 1984. *Dominance and Affection: The Making of Pets*. Yale: Yale University Press.

von Uexküll, J. J. 1992 [1934]. A stroll through the worlds of animals and men: A picture book of invisible worlds. *Semiotica* 89 (4): 319–391.

Valtonen, T. 2016. "Báiki báikki manjis" – "paikka paikan perään". Paikannimet Käsivarren saamelaisessa kulttuurimaisemassa." In T. Nykänen & L. Valkeapää (eds.) *Kilpisjärven poliittinen luonto*: 41–80. Suomalaisen kirjallisuuden seuran toimituksia 1422.

Williams, R., Taylor, G. & Orr, C. 2020. pXRF method development for elemental analysis of archaeological soil. *Archaeometry* 62: 1145–1163.

Wolch, J. R., West, K. & Gaines T. E. 1995. Transspecies Urban Theory. *Environment and Planning D: Society and Space* 13 (6): 735–760.

Yates, T. 1989. Habitus and social space: some suggestions about meaning in the Saami (Lapp) tent ca. 1700–1900. In I. Hodder (ed.) *The Meaning of Things. Material Culture and Symbolic Expression*: 249–262.

Archaeological Discovery of Medieval Vyborg. The Town Development in the 15th-17th Centuries According to the Materials Obtained during the Archaeological Excavations in 1998-2012

Aleksandr I. Saksa

Aleksandr I. Saksa, Institute for the History of Material Culture, Russian Academy of Sciences, Dvortsovaya emb., 18, 191186 Saint-Petersburg, Russia, e-mail: al-saksa@yandex.ru

Abstract

The Institute for the History of Material Culture of the Russian Academy of Sciences (IIMK RAN, St. Petersburg) carried out rescue and research excavations in different parts of the historical centre of Vyborg in 1998-2013. The Karjalaisen Kulttuurin Edistämissäätiö (KKES) foundation played a major role in the success of the excavations on account of its part in funding the work. The second essential matter was that we used local young people as the workforce from the very beginning, and a significant share of them continued working in the following years as well. This played a major role in improving the viewpoints of local people, because it demonstrated how valuable the cultural layers (wooden constructions, artefact finds) of Vyborg are.

The peculiarity of the original bedrock surface with its hollows and hills was already clear in the beginning of the research. As there was little room for construction activities on the slopes of the central town hill, the builders of the town had to fill the rock hollows with the cultural layers of previous times (fires were a regular occurrence in the city) and leftover wood chips from intensive construction activities. There is valuable information about the early stages of the history of Vyborg; in all likelihood, the town cape was already occupied before the Swedes arrived. What seems to have happened is that this native population was absorbed into the population of the emerging town. The natives' main livelihoods were fishing and animal husbandry, and their products happened to be in high demand within the castle and also among the builders of the city wall. Excavations of the plot and in other parts of the town have revealed no finds dating back further than 1410. The excavations carried out so far show that the town, established on the cape, was destroyed so thoroughly by the Novgorodians in 1411 that it had to be rebuilt from the ground up.

1 Introduction

The excavations carried out in 1998-2012 in Vyborg by the Vyborg archaeological expedition of the Institute of the History of Material Culture of the Russian Academy of Sciences (IIMK RAN) and headed by the author have been described in a number of articles and talks given at scientific conferences and symposia (Saksa 2008: 510-514; Saksa 2009a: 226-239; 2009b: 152-162; 2009c: 139–146; 2009d: 473-484; 2010: 238-252; 2011: 105-113; 2015: 256–271; 2016: 246–261; 2017a: 511-524; 2017b: 66-80; 2018a: 65-80; 2018b:

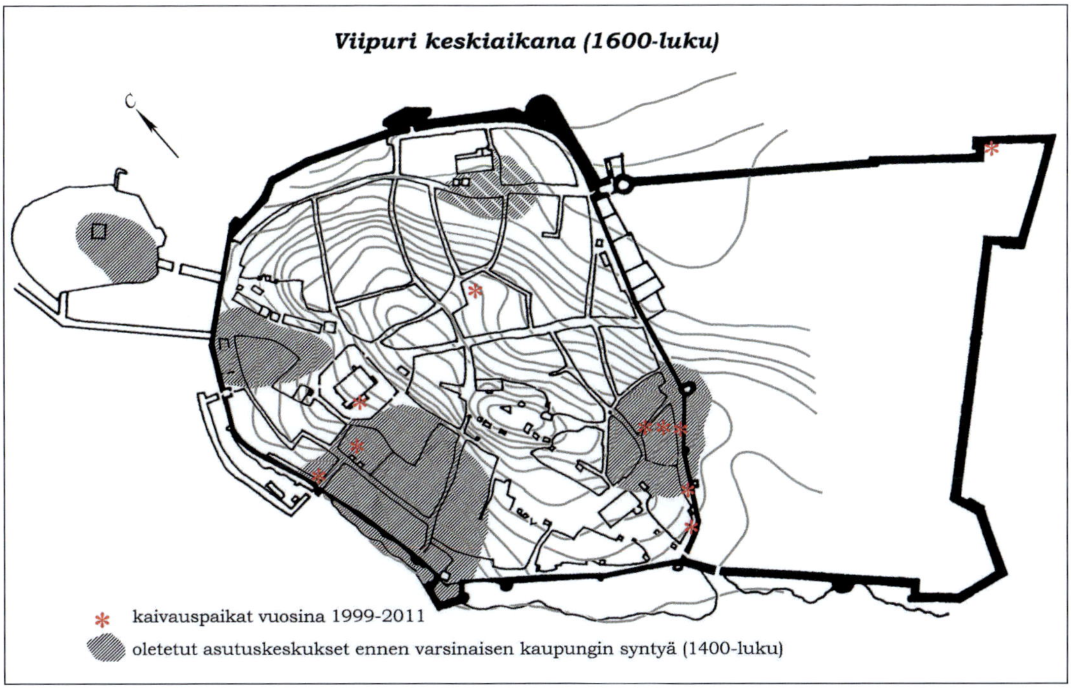

Figure 1. Map dated to the 1630s. The excavation areas of 1999-2011 are marked with red stars, The areas marked with grey are possible locations of medieval settlement.

198-213; 2019: 348–378; Saksa & Taavitsainen 2008: 393-400; Saksa et al. 2002a: 4–10; 2002b: 37–64; 2020: 12–37) (Fig. 1).

This article aims to describe the history of our excavations, the conditions under which they were carried out and how the idea of the early history of Vyborg gradually changed and became more precise as new materials came to light. The research situation after 1998 differed from the earlier one substantially because the relatively extensive rescue and research excavations were carried out for the entirety of the Old City. Now, the excavations extended all the way to the surface of the bedrock, whereas previously this had not necessarily been the case.

2 The new stage of excavations in 1998-2000

A new stage of the archaeological studies of Vyborg began in 1998 with a touch of drama. IIMK RAN received a message from Vyborg that the medieval Ratushnaya Tower (Town Hall Tower – Raatitorni) with the adjacent land plot had been transferred to Katakombnaya Church, to be used for the church's own purposes.

Soviet archaeologist V. A. Tyulenev (Tyulenev 1995: 47-48) had conducted excavations on the same plot in 1978-1994. He discovered the medieval town wall on the foot of the tower on the southwestern side. The wall was dated to the 1470s, because Vyborg Castle was ruled at that time by Erik Axelsson Tott, who is known for the construction of Olavinlinna Castle at the eastern border of the Swedish realm, which occurred at the same time as that of the town wall of Vyborg. To enable the construction of the room / rooms by Katakombnaya Church, a pit was dug, and as a result a part of the stone wall adjacent to the tower, four metres long and two metres deep, was revealed.

IIMK RAN needed to react very quickly, so under these circumstances which allowed

Figure 2. The foot of the stone town wall near the Ratushnaya Tower dated to the 1470s. Photo A. Saksa.

no delay, we turned to Karjalaisen Kulttuurin Edistämissäätiö (KKES) (Foundation for Karelian Culture; Joensuu, Finland) for assistance, because it was impossible at that time to quickly raise even a modest sum of money to carry out excavations in Vyborg. Upon receiving a positive reply, we, together with our colleagues V. A. Nazarenko and V. Ya. Shumkin, immediately came to the city in order to clear the revealed part of the wall, to study how the wall was constructed (stones, mortar), and to conclusively prove that it was the medieval town wall.

The pit widening works were suspended, and in 1999 and 2000 we carried out new excavations at that site with the financial support of KKES (Saksa 2010: 238-252).

The excavations in 1998-2000 resulted in new information about the technical peculiarities of the stone wall and tower's construction. When constructing the town fortification facilities, the builders had been faced with a problem regarding the very uneven bedrock surface. The Ratushnaya Tower itself was in a place where the rock surface is less sloping.

When the construction of the fortification facilities began, these peculiarities of the terrain in Vyborg led to an unusual solution in order to make the wall stable: in the place where the southern curtain wall met the tower, three lower stone layers laid at the foot of the tower were attached to the outer surface of the section of the tower that had been already constructed, and the upper stones were laid down simultaneously as the Ratushnaya Tower was being constructed and were bonded with the stones of its southern wall (Fig. 2).

Despite all these precautions, a hundred years later the wall was in such poor condition that the construction of bulwarks, which were fortification facilities made of log structures filled with sand and stones and covered in soil (Kauppi & Miltšik 1993: 38, 138), was needed in order to support the wall and keep it functioning. This was done in accordance with the royal decree of King John III of Sweden in 1571. The remains of the bulwarks were found near the outer side of the wall during the excavations in 1999-2000.

The studies of the stone town wall have led to a new question which is important for understanding the history of the early days of the town and is related to the development of Vyborg in the second half of the 15th century. Now that we know that the wall was constructed while taking the peculiarities of the on-site terrain into account, a question of intent has arisen. Namely, if the wall was built with future town development in mind, with the use of all available ground for construction being considered, or if there were already existing town dwellings on the construction spot, with the wall leaving a few houses or plots outside the town area.

An answer to this question was found during the excavations. A small excavation area at the internal side of the wall near the Ratushnaya Tower showed that the lowermost stones of the town wall were covered with a 30-cm thick cultural layer on the bedrock. After it was removed, it turned out that there was some mortar between the stones, which means that the lower part of the wall had been exposed to open air and rain for only a short time. In other words, the cultural layer had formed shortly after the construction of the wall.

In 2000, we received word from Vyborg that a new residential building was being planned at the corner of Ul. Storozhevoy Bashni street (Fi. Vahtitorninkatu) and Ul. Titova street (Fi. Possenkatu). That same year, with the financial support of KKES, we opened a small excavation area on the town wall line (No. 4) and two survey trenches to study the cultural layer on the construction site (Nos. 3 & 5) (Fig. 3). The medieval town wall runs from the Ratushnaya Tower to the Market Place. The remains of the wall had been discovered in front of the Round Tower by Tyulenev (his results were published posthumously in 1995).

Survey Trench No. 3 was aimed at getting additional information about the stone town wall of the 1470s. A part of the wall revealed in the excavation area stood on the rock brow and swung slightly to the internal part of the town, which stressed the abovementioned necessity for the builders to consider the topography of the terrain and use as much of it as possible.

In Excavation Area No. 4 (24 sq.m), the Vyborg expedition found well-preserved cultural layers reflecting the periods from the moment of development of the eastern town area (the 1470s) to a stone-paved street on the spot before World War II. In total, we studied eight diverse horizons with foundations of wooden structures. It was the first evidence that material remains of earlier times lying beneath the modern ground surface are very well preserved in the historical part of Vyborg.

In the lowermost horizon, on the bedrock, there was a stone oven and a water pipe made of wooden blanks wrapped in birch bark. Unfortunately, only a part of the structure could be seen in the survey trench.

There were wooden remains of a building that had an oven and a water pipe. Two building logs have been determined to date from 1478 and 1480 by dendrochronology (Pentti Zetterberg pers. comm.). This structure, which was most likely a sauna, was built shortly after the completion of the stone town wall.

In Survey Trench No. 5 (32 sq.m.), there was only a stone paved street and stone foundation located at its northern part. This street probably ran from the Tower of St. Andreas to the Dominican Convent.

3 The excavations in 2001-2003

Part of Trench No. 5 got into the excavation area of 2001, which became the first large excavation area (92 sq.m.) in the history of the studies of the cultural layers in inland Vyborg. The thickness of the cultural layers under the modern ground surface reached 2,75-3,25 metres, which included five separate horizons with the structure remains. The excavation area was set up between Excava-

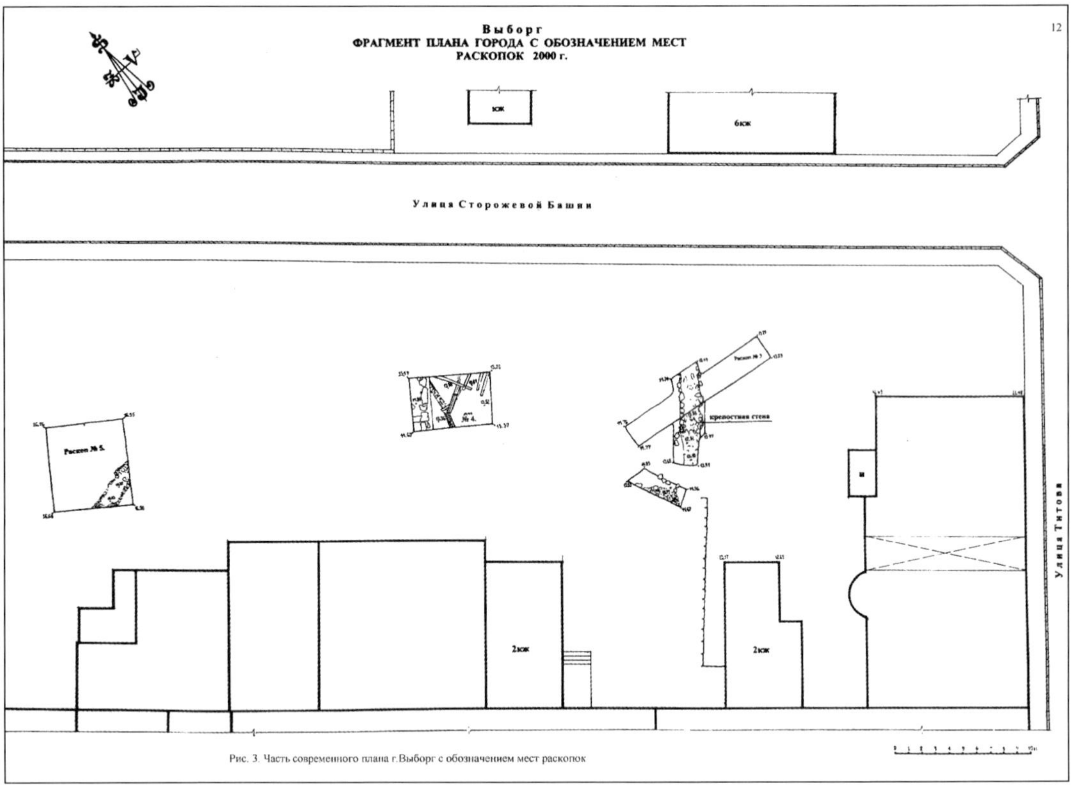

Рис. 3. Часть современного плана г.Выборг с обозначением мест раскопок

Figure 3. Map of the excavation areas in 2000. Drawing Larisa Korchagina.

tion Area No. 4 and Trench No. 5 of 2000, and it was the only free space left during the construction of a block of flats.

The excavations were carried out at the corner of Ul. Storozhevoy Bashni street and Ul. Titova street in 2000-2002 for a total area of 145 sq.m.

The aim of the excavations in 2001 was to study the history of development of the eastern part of the medieval town area and dynamics of the town expansion. In the lowermost part of the excavation area on the rock we found floor planks of the building and wooden barrels, probably parts of a water pipe system. Radiocarbon analysis from the birch bark covering of a barrel dated it back to 1245-1283 (Saksa et al. 2003: 15–20; 2020: 16).

This discovery means that, along with the fortified settlement on the Castle Island and the medieval town facing the castle, there could have been small onshore Karelian settlements before the Swedes came and established Vyborg Castle in 1293.

The following horizon (2) was represented with re-used log structures which formed a flat platform for houses and other buildings. The remaining parts of these dwellings and household buildings dated to the second half of the 16th century. They occupied the entire space of the excavation area. There were wooden buildings on both sides of a street. Part of the street was covered by planks, while another part was paved with stones. The street ran from west to east (Fig. 4) (Saksa 2009a: 228-232; 2009b: 154-158; 2017b: 66-80).

The gap between dating the lowermost horizon to the second half of the 13th century and dating the following horizon with already dense settlement to the second half of the 16th century could be explained by an

Figure 4. Excavation Area of 2001, Horizon 2. Photo A. Saksa.

interruption in settlement formation in the eastern part of Vyborg. This break could have been caused when Novgorodian forces attacked Vyborg in 1411 and burnt down the town (castle) suburbs.

Similarly in the central part of the town, in excavations in 2006 at Ul. Vyborgskaya street 8 (Fi. Luostarinkatu 8) only 120 metres from the town church the earlier horizon was dated back to the 1420s, while in the western seaside part of the city in excavations at Yuzhny Val (Fi. Etelävalli) in 2003 the horizon was dated back to the 1410s. This fact leads to a conclusion that the town, after its complete destruction by the Novgorodians in 1411, was actually built anew (Saksa 2015: 256-271; 2017b: 66-80).

For a long period, this eastern part of the town was unsafe, and it developed quickly after the stone town wall was built in the 1470s and gave it the necessary protection. Afterwards, within the framework of the wider construction activities which, as revealed during our further excavations, covered the whole town, the eastern part, close to the town wall where the sloping hill required large-scale flattening work for future buildings, including the construction of the underlying log platforms, was developed. This kind of undertaking, aimed at filling rock hollows with mineral soil, sand and debris such as wooden chips, was carried out almost throughout the whole medieval town of Vyborg before houses and household buildings were erected.

Here, in the easternmost part of the medieval town area, in Excavation Area No.6 which was 92 sq.m. (13x7 metres) in size, the difference in bedrock levels reached 0,5-1,1 metres, and throughout the whole area studied in 2000-2002 between Sur-

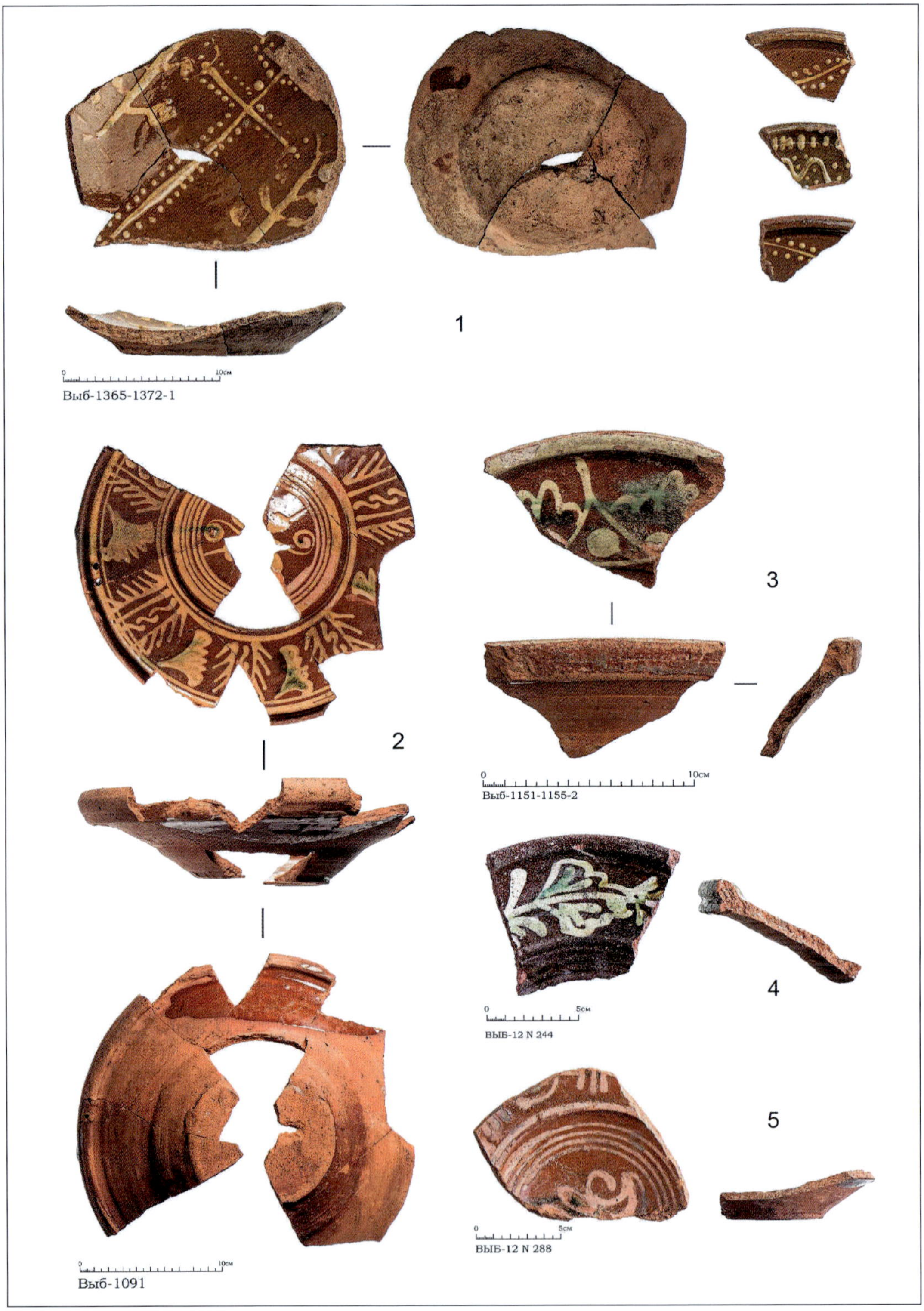

Figure 5. Pottery of the 16th and 17th centuries. Photo Ivan Samovarov.

vey Trench No.4 ('the sauna' of Horizon 8 – 13,56-13,90 metres above sea level) and Survey Trench No.5 (the stone paved street, 15,03 metres above sea level) the difference was even more considerable.

The street revealed in Horizon 2 of Excavation Area 6 of 2001 (13,65 – 13,88 metres above sea level) continued in Survey Trench No. 5 of 2000. The distance between excavation areas 5 and 6 was only 7,5 m. This means that the bedrock surface rises over a metre at such a short distance.

This difference in comparison with the level of the street laid in the second half of the 16th century can be explained by the fact that the excavation area of 2001 reached the lowering of the rock surface, which had 20-40 cm of sand in the bottom and was filled with a 50-80 cm thick cultural layer consisting of huge amounts of wood chips, livestock manure and domestic animal bones. Bark floats for a fishing net made of birch bark and a stone, as well as coins dated to the fourth quarter of the 15th – 16th centuries, were discovered. The coins are from Dorpat and Reval, and a few of them are Swedish.

Local redware pottery and German stoneware, metal, leather, wooden and birch bark finds as well as their shreds only appeared between the 16th and 17th centuries in the 2nd and 3rd horizons.

Rainwater was a problem in medieval times, and water collection barrels, dated to the 13th century, buried in the ground point to this issue. German stoneware of the 16th century and red clay pottery from Western Europe, such as pipkins (three-foot pots), bowls, jugs and ornate table dishes of the 16th and 17th centuries, mark the moment when this territory became a part of the already existing town (Fig. 5).

The dating of this period, which preceded the regular re-construction of Vyborg started in the 1640s, when the town plan was changed into rectangular plots and straight-running streets ca. 1640, refers to early 17th century layers where the logs (walls of wooden houses) did not follow the post-medieval rectangular town plan, which is proven by the Swedish coins of 1634-1640 found here (Horizon 4 of 2000 and Horizon 1 of 2001).

In 2002, construction of a residential building began on the above-mentioned plot, and it became impossible to proceed with the excavations. At the same time, however, the results of the first excavations carried out by us in the city revealed the well-preserved and rich cultural layer embracing many centuries and containing remnants of wooden structures and a considerable amount of finds and showed the town's rank among other ancient towns of the region, its extraordinary value as well as the scope of the losses already borne. The most important thing needed for the preservation of the city's archaeological heritage, the establishment of protection zones in Vyborg, has not yet been done. However, in 2016 the restoration concept of the historic centre of Vyborg, including the castle, was approved by the President of the Russian Federation, and it was financed by the Ministry of Cultural Affairs and the government of Leningrad Oblast. The approval was influenced by the results of the archaeological excavations.

Excavations in 2003 were carried out as protective measures at the customer's expense in the western part of the medieval town near the harbour in Yuzhniy Val street, where a sailors' accommodation centre was being planned. The total area of the excavation site was 60 sq.m, and seven horizons were detected in a 3.5-m thick cultural layer (Saksa 2009d: 473-484).

A radiocarbon sample taken from a plank from the house floor of the lower Horizon No.7 was dated to the 1410s. Below it, in the bedrock beneath the topsoil, seeds of white goosefoot (*Chenopodium album*) dating to approximately 1310 were found (Saarnisto & Saksa 2004a: 259–261; 2004b: 37–42; Saksa 2009d: 484; Saksa et al. 2020: 16). All the cultural layer horizons related to the history of Vyborg up to the 18th century were presented in this excavation area.

Horizon No. 2 was dated to the period of the town being besieged by the army of Peter the Great in 1710, which is proved by the

Figure 6. Excavation Areas of 2004 and 2006 at 8 Ul. Vyborgskaya street. The foundation of the building dated to the first half of the 15th century. Drawing Larisa Korchagina.

burnt layer, cannonballs and their fragments, as well as by other finds.

The lower horizon also showed a burnt layer which could be compared with the large town fires of 1627-1628. Below this layer were cultural layers dating back to the Middle Ages, which preserved the foundations of the small log cabin built repeatedly in the same place.

4 The excavations in 2004 and afterwards

The biggest excavation works for research purposes were fulfilled by us in 2004, 2006, 2011 and 2012 in the city centre, 120 m from the cathedral ruins (a town church until 1554), at the site near 8 Ul. Vyborgskaya street (a so-called guild house). Financial support for the excavations was granted by KKES and Karjalan kulttuurirahasto (Foundation for Karelian Culture), as well as Gazprom Neft, JSC.

The total excavation area was 310 sq.m and the cultural layer was 3-3.4 metres thick. It had ten horizons with well-preserved lower parts of wooden structures like foundations, floors, the lowermost logs of the walls of buildings, paved streets, wells, elements of a drainage system, and different finds (Saksa 2009a: 232-237; 2011: 105-113; 2014: 323-327; 2016: 256-257; 2017a: 511-524; Saksa et al. 2020: 16, 24).

The lower parts of the wooden structures were very well preserved due to intensive construction, when the medieval building remains lay side by side and on top of each other, resulting in the formation of thick layers of wood chips, which allowed us to detect six construction stages on this medieval boundary site, a town at the border of the Swedish kingdom. The earliest wooden structure was a water channel in the excavation area of 2004, which was made of a plank-covered half-log buried in the sand on the mainland and dating to between the 1410s and 1440s (Saarnisto & Saksa 2004a: 259-261; Saksa 2009a: 233, Fig. 10; 2011: 109-110; 2017a: 512-513; Saksa et al. 2020: 16, 24). According to radiocarbon analysis of the seeds of cultivated plants in the underlying layer, the construction works on this site began in the 1310s, which means that agricultural activities preceded the town development (Saksa et al. 2020: 16-17).

5 Summary of the results from Vyborgskaya 8

The materials obtained from the excavations show the development of building construction in the town and the citizens' material culture and trade relations as well as changes in living conditions from the 15th century to the early 18th century. Since the results of these excavations have been published in a range of articles, we will only give a brief summary here.

Construction activities in central Vyborg, to the west from the central town hill which is now a flatland, started in the 15th century and were limited by the considerably irregular surface of the natural terrain (Fig. 1, 6). There was insufficient flat surface for building houses, which is why the first houses, with an area of 10 sq.m, were erected on small sites between the rocks.

At the first stage, the inhabitants' livelihood was based on fishing and livestock keeping. In the lower horizons of the cultural layer we found a lot of floats and net weights indicating this, as well as many bones of domestic animals, which can be explained by the fact that the inhabitants of medieval Vyborg were probably responsible for providing food to the garrison and the castle dwellers (Kasparov 2017: 279-290; Saksa 2008: 510-514; 2009c: 139-146).

Recurring fires and fortification work, which required food supplies and draught force for construction workers, also influenced the life of the inhabitants and the development of the town and left their traces. As a result, sheds and horse stables were a big part of household buildings in the town, and the earliest ones were built on Vyborgskaya Ul. 8 right after the disastrous city fire of 1477.

Figure 7. Excavation Area of 2012 at 8 Ul. Vyborgskaya street. The foundation of the building dated to the second half of the 15th century. Photo A. Saksa.

Figure 8. Commercial seals. Photo Ivan Samovarov.

Considerable changes in construction took place in the 15th century, when the bedrock which divided the site into two sections was covered with the cultural layer and the site became a single built-up area. On this site there was a residential building and household buildings, and a new row of buildings was drawn. The peculiarity of this boundary site was a Holy Spirit guild house located on it, which had some impact on the nature of the finds. The finds were commercial seals, book clasps, coins (the earliest of these were minted in the 1420s), locks, kitchen knives and ornamented kitchenware samples, as well as footwear and fragments, and all those things complied with the 'commercial trademarks' widespread at the time in the cities of Western Europe and the south coast of the Baltic Sea (Fig. 8).

Vyborg in the 15th and 16th centuries became an integral part of the medieval urban civilisation of the Baltic region on its eastern edge. The only thing that differentiated Vyborg from these coastal historic cities, from Lübeck to Riga, Tallinn and Turku, is the fact that Vyborg, which had come into existence relatively late, got all these benefits of civilisation ready for use, while other cities needed hundreds of years to achieve the required level of urban culture.

The second difference is that the town structure in Vyborg was forming in a different manner, unlike other cities of the region, which was caused by a complicated initial terrain for the future town area. It was formed in spots, in separate parts – on limited areas available for construction works until all the hollows in the initial surface were filled with the leftovers of construction activity and it became possible to establish blocks with a connecting street network.

The deceased in the town cemetery, discovered in 2012, were buried near the southern wall of the medieval town church, right near the thick stone foundation of the church. Here, in the sand on the bedrock, medieval burials were found, and the earliest ones are dated to the second half of the 14th – early

15th century (Saksa 2014: 261-270; Saksa & Savolainen 2015: 53-70; Saksa et al. 2017: 167-189; 2020: 16). The construction of the stone church, preceded by the wooden one, is now dated to the 1430s (Hiekkanen 2007: 548-551). These are new excavations that can shed light on this issue.

According to the excavation materials of 1998-2012, there are reasons to suppose that, along with the fortified settlement on the Castle Island on the peninsula, there could have been one or several Karelian seaside villages consisting of a few households predating the founding of the town, with their inhabitants later becoming absorbed by the town (see Tyulenev 1995: 17, 18). It should be noted that the sea level was about 1.5 metres higher than now and the coastline was different.

With the recent excavation results taken into account, J.W. Ruuth's conclusion seems to be correct, i.e. that the first inhabitants of Vyborg moved from the Castle Island which had become too small for them, and in the first half of the 14th century they set their houses in a horseshoe shape on the shore of the town cape opposite the castle (Niukkanen et al. 2014: 51, 55; Ruuth 1982: 47-48).

At the same time, when Borchardt's / Weckrooth's house lying right opposite the castle was reconstructed in the first decade of the 21st century and the soil was temporarily removed from the top of the rock surface, it turned out that the rock in this place inclined considerably to the narrow straight between the Castle Island and the mainland and was very uneven, which was quite typical. Thus, we can suggest that the initial site of the fledgling town of Vyborg neighboured the town church area in the northwestern part of the peninsula, to the west and southwest from the hilltop.

Our excavations in 2003 in Ul. Yuzhniy Val street showed the earliest cultural layer there was dated to the 14th century and the

Figure 9. Excavation Area of 2007 at 4 Ul. Krasnoflotskaya street. Photo A. Saksa.

first log cabin built on this site was dated to the 1410s (Saarnisto & Saksa 2004a: 259-261; Saksa 2009a: 232-238; 2009d: 484).

This peculiarity of the terrain in Vyborg defined the following construction activities in the town. At the beginning of the Modern Age there were some undeveloped sites on the hilltop in the central part of the peninsula, and construction only became possible there after stone masonry had become widespread in the second half of the 17th century, which was clearly shown by our excavations of 2007 in Ul. Krasnoflotskaya street (Harmaidenveljestenkatu). Here, on the eastern slope of the hilltop, the rock had a 'wave-like' surface with considerable elevation differences (Saksa 2009a: 237-238, Fig. 16; 2015: 257; 2016: 259-260). On this site, within the framework of the protection works, the rescue excavations were carried out and we cleared the cellars in the house dated to the late 17th century – 18th century which got an additional building in the early 19th century (Fig.9).

References

Hiekkanen, M. 2007. *Suomen keskiajan kivikirkot*. Helsinki: Suomalaisen Kirjallisuuden Seura.

Kasparov A. K. 2017. The economy of medieval and post-medieval Vyborg, Russia, in its historical context. In Rowley-Conwy P., Serjeantson D & Halstead P. (eds) *Economic zoarchaeology: Studies in Hunting, Herding and Early Agriculture*: 279-290. Oxbow Books 2017.

Kauppi, U.-R. & Miltšik, M. 1993. *Viipuri: Vanhan Suomen pääkaupunki*. Suomalaisen Kirjallisuuden Seuran toimituksia 593.

Niukkanen, M., Seppänen, L. & Suhonen, M. 2014. Kaupunkirakentaminen Suomessa keskiajalla. In H. Lilius & P. Kärki (eds.): *Suomen kaupunkirakentamisen historia*: 1 osa: 28-94. Helsinki: Suomalaisen Kirjallisuuden Seura.

Ruuth, J. W. 1982. *Viipurin kaupungin historia: 1osa: vuoteen 1617*. Helsinki: Torkkelin säätiö. R. Rosen et al., revised edition.

Saarnisto, M. & Saksa, A. 2004a. Radiohiiliajoituksia Viipurin arkeologisilta kaivauksilta. In J. Korpela *Viipurin linnaläänin synty. Viipurin läänin historia* II: 259-261. Jyväskylä.

Saarnisto, M. & Saksa, A. 2004b. Radiocarbon dates from archaeological excavations in Viipuri. The corner site of former Uudenportinkatu and Etelävalli. *Fennoscandia archaeologica* XXI (2004): 37–42.

Saksa 2008 = Сакса А. И. 2008. Древний Выборг. Становление средневекового города. *Труды II (XVIII) всероссийского археологического съезда в Суздале*: 510-514, Том II. М.

Saksa, A. 2009a. Kaupunkiarkeologia Viipurissa. Kaupungin alkuvaiheen ongelma. In H.-M. Pellinen (ed.) *Maasta, kivestä ja hengestä. Markus Hiekkanen Festschrift*: 226-239. Saarijärvi.

Saksa, A. 2009b. Baugeschichte und kulturhistorische Stratigraphie des mittelalterlichen Wyborgs (nach Materialien der Ausgrabungen 1998-2008). In A. Caune & I. Ose (eds) *The Hansa Town Riga as mediator between East and West: Proceedings of an international scientific conference dedicated to 70 years of archaeological research in Riga, Held in Riga, Latvia, on 23-25 September 2008*: 152-162. Riga: Institute of the History of Latvia Publishers.

Saksa 2009c = Сакса А. И. 2009. Средневековый Выборг: начало городской истории (по итогам раскопок 1999–2006 гг.). In Е. Н. Носов (ed) *Археология и история Литвы и Северо-Запада России в раннем и позднем средневековье.Доклады Международного российско-литовского семинара. Санкт-Петербург, 4–8 декабря 2006 г.*: 139-146. СПб.: Нестор-История.

Saksa 2009d = Сакса А.И. 2009. Раскоп на ул. Южный Вал в Выборге. Стратиграфия культурных напластований средневекового города. In А. Е. Мусин (ed) *Хорошие дни. Памяти Александра Степановича Хорошева*: 473-484. Великий Новгород. Санкт-Петербург & Москва.

Saksa 2010 = Сакса А.И. 2010. Новые данные по городской стене Выборга 1470-х гг. In Е.Н. Носов, С.В. Белецкий (ed) *Краеугольный камень. Археология, история, искусство, кльтура России и сопредельных территорий. 80-летию со дня рождения Анатолия Николаевича Кирпичникова посвящается*: 238-253. Том II. Москва. «Ломоносовъ».

Saksa 2011 = Сакса А.И. 2011. Раскоп на ул. Выборгская, 8. К проблеме ранней истории средневекового Выборга. In П.Г. Гайдуков (ed) *Археология и история Пскова и Псковской земли. Семинар имени академика В.В. Седова. Материалы 56-го заседания (7-9 апреля 2010 г.)*: 105-113. Москва-Псков: Институт археологии РАН.

Saksa 2014 = Сакса А.И. 2014. Старый кафедральный собор в Выборге – уникальный объект культурного наследия. In Марк Е. Ткачук (ed) *Время войны. Победители и побежденные. Ad gloriam. А. Н. Кирпичников*: 261-270. Stratum plus. № 6. 2014. Санкт-Петербург, Кишинев, Одесса, Бухарест.

Saksa 2015 = Сакса А. И. 2015. Средневековый Выборг как восточное продолжение городской цивилизации побережья Балтики, In Л.А. Беляев, П.Г. Гайдуков (eds) *Города и веси Средневековой Руси. Археология, история, культура. К 60-летию Николая Андреевича Макарова*: 256-271. Москва – Вологда. «Древности Севера».

Saksa 2016. The early stages of the history of Vyborg: The results of Archaeological Research 1998–2012. In Pirjo Uino & Kerkko Nordqist. (eds) *New Sites, new Methods. Proceedings of the Finnish-Russian Archaeological Symposium. Helsinki, 19-21 November, 2014*: 246-261. Suomen Muinaismuistoyhdistys ty – Finska Fornminnesföreningen rf – The Finnish Antiquarian Society. Iskos 21, 2016. Helsinki.

Saksa 2017a = Сакса А.И. 2017. «Дом купеческой гильдии» (Гильдия Святого Духа) в Выборге: к вопросу о локализации. In Мусин А.Е. (ed) *В камне и бронзе. Сборник статей в честь Анны Песковой*: 511-524, Труды ИИМК РАН. Т. XLVIII – СПб.: ИИМК РАН, ООО «Невская Книжная Типографии». Abstract: Saksa A.I. "House of Merchant's Guild" (Guild of the Saint Spirit) in Viborg. Problem of Localization. P. 645.

Saksa 2017b = Сакса А.И. 2017. Средневековый Выборг. Восточная часть города по археологическим данным. По материалам раскопок 2000-2001 гг. In Е.Н. Носов (ed) *Новгород и Новгородская земля. История и археология. Материалы XXXI научной конференции, посвященной 85-летию археологического изучения Новгорода*: 66-80. Великий Новгород, 25-27 января 2017 г. Выпуск 31. Великий Новгород.

Saksa 2018a. Viipuri 1500-luvun alkupuolella eli millaiselta kaupunki näytti Mikael Agricolan ajan katutasolta. *Suomen keskiajan arkeologian seura (SKAS). 2/2018*: 65-80. Helsinki.

Saksa 2018b = Сакса А.И. 2018. Выборг в конце эпохи Средневековья – начале Нового времени. In В.К. Сингх (ed) *Нескончаемое лето. Сборник статей в честь Елены Александровны Рыбиной*: 198-213. Москва – Великий Новгород – Санкт-Петербург: Любавич.

Saksa 2019 = Сакса А. И. 2019. Выборг — город на перекрестке истории. In Ю.А. Виноградов, С.А. Васильев & К.Н. Степанова (eds) *Прошлое человечества в трудах петербургских археологов на рубеже тысячелетий (К 100-летию создания российской академической археологии)*: 348-378. СПб.: Петербургское Востоковедение.

Saksa A. & Savolainen, P. 2015. The Old Cathedral of Viborg. Historical Archaeology of the Easternmost Cathedral of the Swedish Empire. *Mirator 16:1/2015. Papers from the conference on church archaelogy in the Baltic Sea region (26-30 august 2013)*: 53-79. Helsinki.

Saksa, A. & Taavitsainen, J.-P. 2008. Viipuri (Sw. Viborg, Ru. Vyborg). In H. Andersson, G. Hansen & I. Øye (eds.) *De første 200 årene – nytt blikk på 27 skandinaviske middelalderbyer*: 393-400. UBAS – Universitetet i Bergen Arkeologiske Skrifter. Nordisk 5. Bergen: Universitetet i Bergen.

Saksa et al. 2002a = Saksa A., Belsky, S. & Suhonen, M. 2002. Kaupunkiarkeologiaa Viipurissa. *Suomen keskiajan arkeologian seura (SKAS) 2/2002*: 4–10.

Saksa et al. 2002b = Saksa A., Belsky, S., Kurbatov, A., Polykova, N. & Suhonen, M. 2002. New archaeological excavations in Viipuri. Results of field investigations of the 1998–2001 seasons and current research problems of urban history: 37-64. *Fennoscandia archaeologica XIX*. Helsinki.

Saksa et al. 2003 = Saksa A., Saarnisto, M. & Taavitsainen, J.-P. 2003. 1200-luvun lopun radiohiiliajoitus Viipurista. *Suomen keskiajan arkeologian seura (SKAS)*. 3/2003: 15–20.

Saksa et al. 2017 = Липатов А. А., Сакса А. И. & Смирнов А. М. 2017. Архитектурно-археологические загадки старого кафедрального собора в Выборге. Некоторые результаты исследований 2013 г. *Вопросы всеобщей истории архитектуры. Выпуск 8 (1/2017)*: 167-189. Москва-Санкт-Петербург.

Saksa et al. 2020 = Saksa, A., Suhonen, M. & Uino, P. 2020. Viipuri maan alla – keskiaikainen karttakuva ja arkeologinen aineisto. In K. Katajala, M. Hietala, M. Niemi, P. Uino, M. Helminen, A. Härkönen & H. Hirvonen (eds): Vyborg underground – Medieval Map Image and Archaeological Finds. *Viipuri – Vyborg. Historiallinen kaupunkikartasto – Historic Towns Atlas. Scandinavian Atlas of Historic Towns. New Series, No 3, Suomi-Finland*. AtlasArt: 12-37. Helsinki.

Tyulenev 1995 = Тюленев В. А. 1995. *Исследования древнего Выборга*. Санкт-Петербург, 1995.

Materiality and Objects: Multi-disciplinary Approaches to Archaeological Material and Contexts

The XVI Finnish-Russian Archaeological Symposium

The symposium 2019 covers all fresh archaeological and multi-disciplinary approaches. Organizers of the symposium are The Finnish Heritage Agency, The Finnish Antiquarian Society and Finlyandsko-rossijskaya rabochaya gruppa po sotrudnichestvu v oblasti arheologii. The venue is the Auditorium of the Häme Castle, Hämeenlinna. After the two working days, Thursday 10th and Friday 11th October, there is a bus excursion to archaeological sites on Saturday 12th October, Working language of the symposium is English.

PROGRAMME

Torstai 10.10.2019 / Thursday, 10th October

9.30 Opening of the Symposium

Juhani Kostet, General Director, The Finnish Heritage Agency, Helsinki

V. A. Lapshin, Director, Institute for the History of Material Culture of the Russian Academy of Sciences (IHMC RAN), St. Petersburg

10.15 SESSION I

Kerkko Nordqvist, University of Helsinki: **Artefacts on the Move. Movement, Mobility and Stone Age Archaeology of North-East Europe**

Evgeniia Tkach, IHMC RAN: **Stone Age Settlements in the Pskov Region: A Review of Old Materials and New Fieldwork**

Cancelled / *Kristiina Mannermaa & Riitta Rainio, University of Helsinki*: **A Needle Case, a Sound Instrument or Something Else? A Worked and Ornamented Swan (*Cygnus sp.*) Ulna from a Late Mesolithic Male Burial (Yuzhniy Oleniy Ostrov, North-West Russia)**

12.00 Lunch in the Castle

13.00 SESSION II

Teemu Mökkönen, Finnish Heritage Agency & Kerkko Nordqvist, University of Helsinki: **The Use of Mineral Raw Materials in Neolithic Northern Ostrobothnia, Finland**

Andrey Gorodilov, IHMC RAN: **Excavations of Burial Cairns near the Village of Bolshoy Bor on the Northern Coast of the Gulf of Finland**

Elisabeth Holmqvist-Sipilä & Anna Wessman, University of Helsinki: **Lead-Isotope and Geochemical Analyses of Copper-Based Metal Artefacts from the Iron Age Water Burial in Levänluhta, Western Finland**

14.45 Coffee break

15.15 SESSION III

Vladimir Lapshin, IHMC RAN: **Old Ladoga: New Finds and Publications**

Natalia Grigorieva, IHMC RAN: **Some Results of a Multi-Disciplinary Approach to the Study of the Staraya Ladoga Archaeological Sites**

Hanna-Leena Puolakka, University of Oulu: **Coins, Knives and Other Finds – Grave Goods in the Transition Period between the Late Iron Age and the Early Medieval Period in Northern Fennoscandia**

Olli Eranti, University of Helsinki: **Macroscopic Analysis of Lithic Grave Finds from Yuzhniy Oleniy Ostrov**

17.30 End of the first working day

18.30 Reception by the City of Hämeenlinna, Venue: Raatihuone (Address: Raatihuoneenkatu 15)

Perjantai 11.10.2019 / Friday, 11th October

10.00 SESSION IV

Volker Heyd, University of Helsinki: **Migrants, terrestrial diet and intensive networking: A different look at the Corded Ware in Finland, the Baltic Countries, and beyond**

Alexander Vybornov, Samara State University of Social Sciences and Education & Marianna Kulkova, Herzen State Pedagogical University & Peter the Great Museum of Anthropology and Ethnography: **A Chronological Framework for the Development of Finno-Ugric Cultures of the Neolithic and Eneolithic of the Northern Kama Region**

Eugen Kolpakov, Anton Murashkin, Aleftina Kiseleva & Vladimir Shumkin, IHMC RAN: **The Oldest Gressbakken-House Site on the Kola Peninsula**

12.00 Lunch in the Castle

13.00 SESSION V

Anton Murashkin & Anna Malutina, IHMC RAN: **Evidence of Metal Use from Bronze Age sites on the Kola Peninsula**

Eugen Kolpakov, Anton Murashkin & Vladimir Shumkin, IHMC RAN: **Archaeological Research in the South-Western Part of the Kola Peninsula in 2017–2019: Initial Results**

Mika Lavento, University of Helsinki: **The Long-Period Dwelling Site of Multavieru at Lake Höytiäinen in Northern Karelia**

14.45 Coffee break

15.15 SESSION VI

Maria Razzak, IHMC RAN: **Archaeological Survey of Islands in the Gulf of Finland in 2019**

Alexander Saksa, IHMC RAN (Presentation in Finnish and Russian): **Kuinka löysimme keskiajan viipurilaiset: arkeologiset kaivaukset 1998–2012 (Развитие Выборга в XV–XVII вв. по материалам археологических раско-пок 1998–2012 гг.)**

Cancelled / *Oula Seitsonen, University of Oulu*: **Geoarchaeological Approaches to the Taskscapes of Early Reindeer Herders in Northern Sápmi**

Petro Pesonen & Ulla Moilanen, University of Turku & Jarkko Saipio, University of Helsinki: **From Primitive Axe to Petersen X – A Typological Artefact Database from Finland**

17.30 End of the second working day

19.00 Banquette / Evening with Friends, Venue: Hämeenlinnan VPK-talo (Address: Palokunnankatu 18)

Lauantai 12.10.2019 / Saturday, 12th October

BUS EXCURSION

8.00 Meeting place: Sokos Hotel Hämeenlinna

Guided visit to the medieval Holy Cross Church in Hattula, the ancient hillfort of Rapola in Valkeakoski, archaeological sites in Janakkala parish and the the Mesolithic site of Brunaberget in Vantaa.

16.00 Arrival at Helsinki (Central Railway Station)

Participants in the 16th Finnish-Russian archaeological symposium in Häme Castle, Hämeenlinna, on 11 October 2019. Photo: P. Halinen.

Aleksandr Saksa and Mika Lavento. Photo: P. Halinen.

Opening of the Symposium and participants of the Symposium including Vladimir Lapshin, Christian Carpelan, Pirjo Uino and Juhani Kostet. Photo: P. Halinen.